Mechanisms and Machine Science

Volume 28

Series editor

Marco Ceccarelli, Cassino, Italy

More information about this series at http://www.springer.com/series/8779

Amitabha Ghosh · Burkhard Corves

Introduction to Micromechanisms and Microactuators

Springer

Amitabha Ghosh
Aerospace Engineering and Applied Mechanics
Indian Institute of Engineering Science and Technology, Shibpur
Howrah, West Bengal
India

Burkhard Corves
RWTH Aachen University
Aachen
Germany

ISSN 2211-0984 ISSN 2211-0992 (electronic)
Mechanisms and Machine Science
ISBN 978-81-322-2143-2 ISBN 978-81-322-2144-9 (eBook)
DOI 10.1007/978-81-322-2144-9

Library of Congress Control Number: 2014958966

Springer New Delhi Heidelberg New York Dordrecht London

Printed on acid-free paper

Springer (India) Pvt. Ltd. is part of Springer Science+Business Media (www.springer.com)

The Alexander von Humboldt Foundation

Preface

Mechanisms have been used since time immemorial for various tasks. Apart from being indispensable for all kinds of mechanization, mechanisms have always fascinated the human mind. Perhaps, in a very rudimentary sense, mechanisms remotely resemble living objects. In the past, mechanism design and inventing new mechanisms were important activities not only for engineers but also for scientists. In fact, the contributions of mathematicians to mechanism theory have really helped this branch of engineering science grow. With the progress in technology, the role of automation and mechanization increased in importance and courses on Theory of Mechanisms and Machines occupied a major share of all Mechanical and Aerospace Engineering curricula.

However, from the second half of the 1970s and early 1980s a kind of gloom descended on the kinematics community. This was primarily because of the rise of microelectronics, programmable logic controllers, and computer control and numerical control (NC) technology. The need for careful and meticulous mechanism synthesis became redundant as any kind of motion and trajectory could be generated with extreme accuracy using multiple servo drive systems controlled by computers. Many universities the world over gradually removed the subjects in the domain of kinematics and mechanism theory from the curricula. Apart from being very accurate the multiple-drive-based NC motion generation has the added advantage of having the ability to quickly change from one geometric characteristic to another without any change in the hardware. This led many kinematicians to think that 'mechanism theory' was destined to become a dead subject.

Fortunately, along with the computer-based Second Industrial Revolution (which was a threat to the continued existence of the subject) came the trend of miniaturization. Though initially this trend was confined to electronic circuits and devices only, soon it spread to other fields. Now it is felt by many that once miniaturization revolution spreads to other fields the process of the Third Industrial Revolution will begin. The main feature behind the forthcoming revolution in technology will be based on miniaturization of three-dimensional machines and devices capable of manipulating material at the micro, nano, and even molecular levels. The pressure of miniaturization increased due to the huge success of miniaturized electromechanical integrated systems and a new branch of technology—Micro Electro—Mechanical Systems (or MEMS)—is emerging as a powerful entity for controlling the world economy. These are primarily lithography-based two-and-half-dimensional miniaturized devices. But the trend is already rearing its head which indicates the bright future for miniaturized machines and mechanisms also. And for the delight of members of the near-extinct community of kinematicians, the subject of mechanism synthesis has once again become important for applications in miniaturized machines and mechanisms. Using multiple-drive-based motion generation became infeasible and the technologists had to once again depend on synthesized mechanisms to generate the desired motion and trajectories as these became the only practical solutions. It can be safely stated that the rebirth of the subject 'kinematics and mechanism theory' has been possible because of the emergence of 'micromechanisms'.

Expectedly, the rebirth of the subject has brought along a number of fundamentally new concepts and paradigm shifts, which influence the basic configuration design, actuation, and fabrication. On many occasions engineers and technologists have started using principles from life science for the functioning of micromechanisms and microactuators. There are major changes from the point of view of the material used and the energy sources employed. Thus, it will be wrong to think that micromechanisms are just the miniaturized versions of their similar macroscopic counterparts.

As it happens with many new emerging branches of science and technology, in the emerging area of microsystems technology also, the subdivisions are not well-classified and considerable nebulosity exists in many definitions and characterizations. Because of continuous scaling down of microelectronic chips, industrial fabrication units started to become obsolete because of their limitations on size capability. These industrial units became useful for fabricating MEMS devices as, unlike microelectronic chips, these devices did not require a very high level of miniaturization. Since MEMS came directly as a product from the obsolete and abandoned microelectronic industry the growth of the subject has been fast and the expansion of its market very rapid. As a result, at present MEMS devices occupy a major part of the microsystems technology. The emergence of micromechanisms has been relatively slow because of their more involved fabrication process. Most often the subject 'microsystem' deals heavily with the discussions of MEMS technology. But it should be noted that the current area of microsystems, on the whole, consists of both MEMS devices (most of which are for sensing) and micromechanisms (which really perform the manipulation of material at the micro, nano, and molecular levels).

This introductory volume deals with primarily the active micro devices—micromechanisms—and the microactuators for driving the mechanisms. The design problems and concepts used in micromechanisms are substantially different from the traditional passive devices for sensing. It is hoped that this volume introducing the rudiments of micromechanisms and microactuators will be helpful to students of this subject and can be used to develop an introductory course.

Up till now, the development of the subject of micromechanisms has been somewhat disorganized in manner. A number of basic aspects are still in their infancy. Apart from this the R&D work on micromechanisms is mostly based on individual problems and specific systems. So, it is not always easy to identify a generalized discussion applicable to the whole class of these systems. In this introductory book, we attempt to identify the important aspects of miniaturization as it is expected to bring revolutionary changes in technology in the years to come. A separate chapter is included to discuss scaling laws as this topic is of utmost importance as far as miniaturization and microsystems are concerned. To make the volume self-contained, we have presented material on the general theory of mechanisms and techniques for synthesis of conventional mechanisms in the introductory chapters. Many micromechanisms are initially designed following the approaches adopted for their macroscopic counterparts. A general introduction to micromechanisms is given in a chapter preceding the chapter on their design. The chapter on design is followed by a chapter presenting some topics on the dynamics of micromechanisms.

The drive systems and actuators for micromechanisms are of extreme importance, as the impact of scaling effects is maximum on the actuation principles. Besides, there are many fundamentally new approaches for actuating micromechanisms. As it may have already become clear to many, when designing micromechanisms, a lot can be learned from the living world. Keeping this in view, similarity with biological systems has been pointed out in sections on design and structure of micromechanisms and microactuators. The chapter on microfabrication also discusses topics like 'self-assembly' that is closely related to life science.

Finally, the text presents the major fabrication techniques and the future possibilities.

The authors thank the Alexander von Humboldt Foundation for the support provided for the exchange visits, which has been the prime mover of this book project.

Amitabha Ghosh
Burkhard Corves

Contents

About the Authors

Amitabha Ghosh did his Bachelor's, Master's, and Doctoral degrees in Mechanical Engineering from Calcutta University in 1962, 1964, and 1969, respectively. After serving as a Lecturer in Mechanical Engineering at his alma mater, Bengal Engineering College, Shibpur (now an Institute of National Importance—Indian Institute of Engineering Science and Technology, Shibpur) from 1965 to 1970 Prof. Ghosh joined the Indian Institute of Technology, Kanpur, in January 1971 as Assistant Professor. Since June 1975 he served at the institute as Professor of Mechanical Engineering till his retirement in 2006. During 1977–1978 Prof. Ghosh visited RWTH Aachen as a Senior Fellow of the Alexander von Humboldt Foundation and, subsequently, he visited RWTH Aachen with Humboldt fellowship many more times till 2012. He served the Indian Institute of Technology, Kharagpur, as Director from 1997 to 2002. His primary areas of research include Manufacturing Science, Robotics, Kinematics and Mechanism Theory, and Dynamics of Mechanical Systems. Professor Ghosh has written a number of textbooks which are very popular both in India and abroad. He has guided many master's and doctoral students and has published a large number of research papers. He received many academic awards including a number of Calcutta University Gold Medals, D.Sc. (h.c.), Distinguished Teacher award from IIT Kanpur and Award for Excellence in Research by the National Academy of Engineering. He is a fellow of all the four national science and engineering academies in India.

Burkhard Corves is currently University Professor and Director of the Department of Mechanism Theory and Machine Dynamics at RWTH Aachen University, Germany. He received his Ph.D. in 1989 from the same university. Dr. Corves has to his credit several books and almost 200 technical papers covering different topics such as robotics, cam mechanism, glass machinery, pneumatic mechanisms, computer-aided graphical analysis and synthesis in mechanism theory, mechatronic development, and machine design strategy. He also actively teaches courses in machine and mechanism design, kinematics, dynamics, and robotics.

Chapter 1
Introduction

Abstract This introductory chapter provides a brief outline of the kinematic analysis and synthesis of plane mechanisms. In MEMS devices, the system's functions mostly depend on elastic deformation of simple elements like cantilevers. However, in case of micromechanisms, a system's operation depends on the coordinated movement of different links like a normal macroscopic mechanism. The only difference lies in the fact that the joints are replaced by flexural hinges. This chapter illustrates the methods of kinematic analysis. A later section presents methods of kinematic synthesis to achieve desired movements. The same methods are in general applicable to mechanisms of microscopic sizes. A brief discussion on actuators is also presented.

Keywords Mechanisms · Kinematic analysis · Dimensional synthesis · Actuators

1.1 Introduction

Since the dawn of civilization, man has been trying to make use of physical laws and principles for accomplishing useful tasks. In many occasions, the effort goes in achieving results which are beyond the capacity of unaided human beings—be it in moving fast, applying large forces, making accurate movements and so on. The idea of using appliances came with such objectives, and the origin of mechanisms and machines can be traced to that. It is known to us very well that one can draw a perfect circle using a compass irrespective of her/his individual drawing skills. A child can break a nut with a crusher irrespective of his/her personal strength. Thus, appliances can enhance the typical human capabilities.

Apart from using mechanical devices for improving and enhancing human physical abilities, devices or appliances had been used to impress common man by exhibiting miraculous phenomena—such as automatic opening of doors. This was the case with the group of people (mostly priests of ancient temples) primarily to command respect of common man.

A. Ghosh and B. Corves, *Introduction to Micromechanisms and Microactuators*,
Mechanisms and Machine Science 28, DOI 10.1007/978-81-322-2144-9_1

All such devices and appliances started being called machines, and many such machines for lifting heavy objects, generating mechanical power from flowing fluids, developing large forces and making warfare were developed since antiquity.

Of late, the trend has been to design and develop machines and devices of extremely small sizes. The reasons for such trends are many as will be discussed later in this volume. However, such small-sized machines and devices need special considerations for their design and development. This volume is an attempt to present a rudimentary introduction to the subject. Though the sizes and linear displacements differ enormously for macroscopic mechanisms and micromechanisms, their basic nature of functioning has considerable similarity. So, an introduction to the basic concepts of mechanisms is desirable before micromechanisms are taken up for discussion.

1.2 Mechanisms and Machines

For understanding micromechanisms and microactuation principles, it is essential to have a clear idea about the basic principles of conventional mechanisms and machines. This section is devoted to some foundational concepts that are essential to design, develop and understand mechanisms and machines.

1.2.1 Definition and Types

Although everyone has a vague idea about mechanisms and machines, a very scientific definition is necessary for the professionals. A mechanism is an agglomeration of several interconnected rigid bodies so as to transfer and transform mechanical movements and motions. It is also to be noted that some definite relationship must exist between the input motion and the resulting output movements. In cases where there is no significant force or power transmitted and the primary objective is to produce a desired movement only, the device is called 'mechanism'. On the other hand, if the objective includes transmission of substantial power doing some useful work, the device is called 'machine'. However, in both cases, mechanical movements of desired characteristics are essential features. It is, thus, clear that all machines must be mechanisms but not the vice versa.

As there exist hundreds of thousands type of mechanisms and machines, it is not a simple task to classify all of them. An attempt can be made for some broad classification at this stage. The most important characteristic that differentiates all mechanisms into two major classes—planar mechanisms and spatial mechanisms—is the spatial nature of motion. In case of plane mechanisms, all points of all the members of the device move in parallel planes. So, it is possible to observe the actual motions of all points in one view. It also ensures that motions of all points are either coplanar or confined to parallel planes. On the other hand, in spatial mechanisms, different points of the device move in a general fashion or, in the simplest case, in planes which are not parallel. Alternatively, planar mechanisms may be called two-dimensional mechanisms and spatial mechanisms may be called three-dimensional mechanisms. It is needless to emphasize that analysis and design of spatial mechanisms are far more complex than those in case of planar mechanisms.

Though members of a mechanism are generally rigid bodies, there are some situations where some members show flexibility which is essential for proper functioning. Such mechanisms are called compliant mechanisms. Compliant mechanisms play a very important role in the field of micromechanisms as will be discussed later.

1.2.2 Kinematic Pairs and Kinematic Chains

To begin with, the bewildering variety and complexity of hundreds and thousands of mechanisms create an impression that it is next to impossible to find any order in the matter. However, if one carefully identifies the basic building blocks of any mechanism, a considerable order can be created in the chaotic complexity. As mechanical motion can be transferred from one body to another by physical contact, only this contact of two bodies constitutes the basic building block of mechanisms and are called 'kinematic pairs'. When the contacting surfaces are of geometrically conforming shapes, the contact is over a surface as shown in Fig. 1.1.

In the case shown in Fig. 1.1a, the surfaces are flat and the relative motion of body 1 with respect to body 2 is along a straight line in case of planar mechanisms (where all objects move in two dimensions only). These kind of kinematic pairs are called 'prismatic pairs'. In the other case, the contacting surfaces are cylindrical in shape with same radii. Thus, the relative motion is circular, and the pair is termed as 'revolute pair'. Both these pairs belong to the same class called 'lower pair'. A major difference between 'prismatic' and 'revolute' pairs needs to be discussed here. Figure 1.2 shows a typical prismatic pair (slides) and a revolute pair (usually a hinge).

In case of slides, the force between the two elements 1 and 2 is the friction force ($=\mu N$) where μ is the coefficient of friction and N is the normal force between the

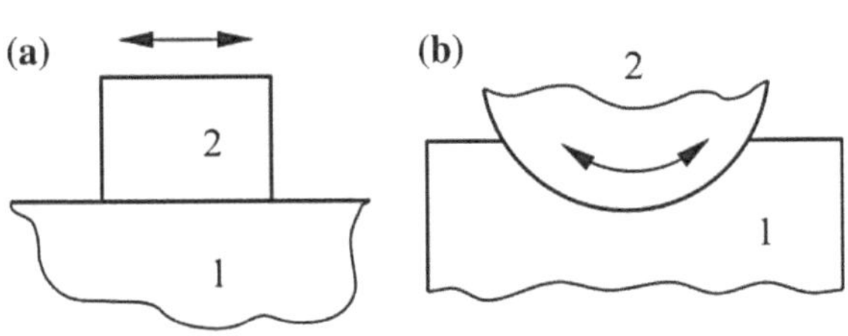

Fig. 1.1 Kinematic pairs with geometrically conforming shapes. **a** Rectilinear movement. **b** Angular movement

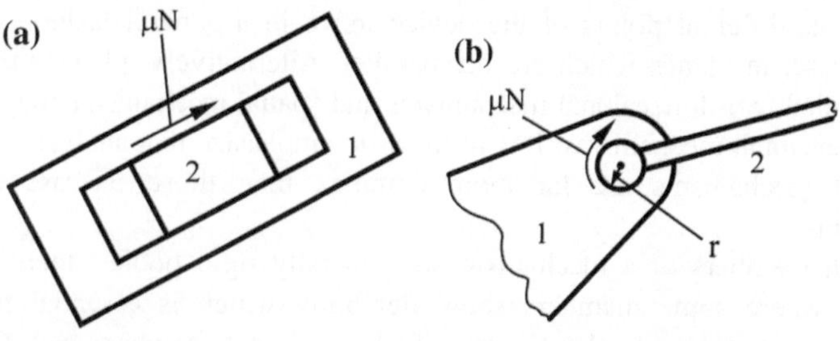

Fig. 1.2 Friction in lower kinematic pairs. a Prismatic pair. b Revolute pair

elements. In the hinge joint, the relative motion is angular and the frictional torque between the two elements is μNr where r is the radius of the cylindrical pin. Since r is typically much smaller compared to other dimensions of a mechanism, the frictional resisting torque is very small and does not impair the relative motion. This is a great advantage, and therefore, hinge joints (i.e. revolute pairs) are generally preferred over prismatic pairs.

The contact between two adjacent elements of a kinematic pair can be between surfaces which do not conform in shape as shown in Fig. 1.3. In such cases, the contact is along a line (or at a point) and not over a surface.

Such kinematic pairs are called ‘higher pairs’. It is readily seen that in case of lower pairs , the degree of freedom of relative motion between the elements is one (when planar mechanisms are considered) and that in case of higher pairs, it is two. Contacting cam and follower and contacting gear teeth are common examples of higher pairs.

A series of kinematic pairs constitutes a kinematic chain as indicated in Fig. 1.4. In cases shown in Fig. 1.4a, b, only lower pairs are used and often kinematicians prefer to call such mechanisms as ‘linkages’.

In Fig. 1.4c, the kinematic chain contains both lower and higher pairs. It should be further noted that a kinematic chain can be either open (as shown in Fig. 1.4a) or closed. A kinematic chain is as such of no use in transferring or transforming motions. For useful work to be done, it is essential to fix one link and this leads to a mechanism. The process of grounding one link of a kinematic chain is called ‘kinematic inversion’. Different inversions of a kinematic chain may give rise to different mechanisms. After inversion, if there is a unique relationship among the

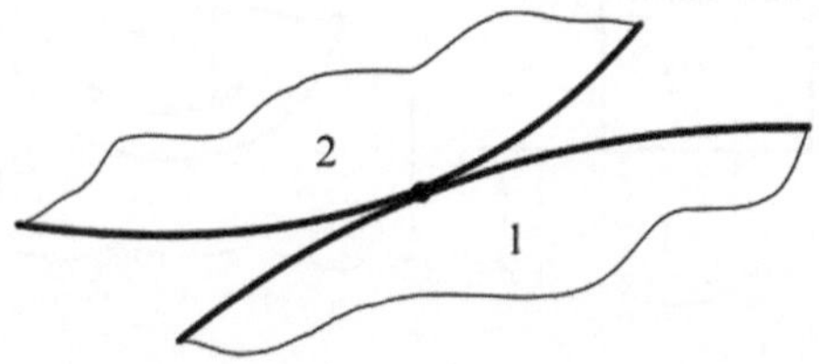

Fig. 1.3 Higher kinematic pair

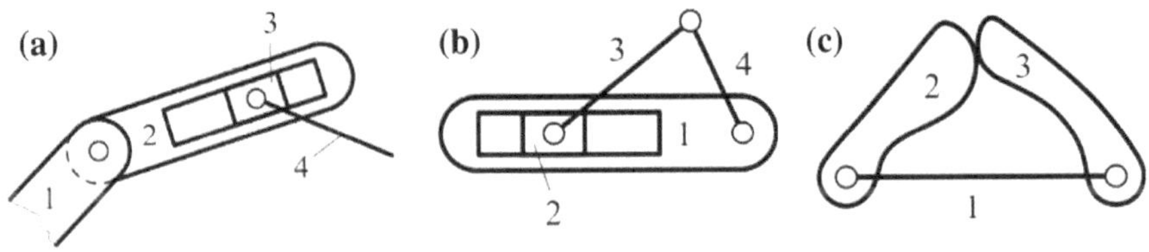

Fig. 1.4 Kinematic chains. **a** Open chain. **b** Closed chain with lower pairs only. **c** Closed chain with lower and higher pairs

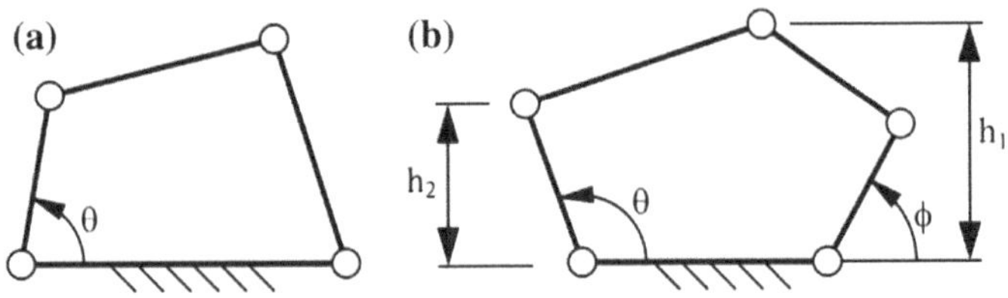

Fig. 1.5 Degrees of freedom of mechanisms. **a** A four-bar mechanism with one degree of freedom. **b** A five-link mechanism with two degrees of freedom

motions of the moving links, i.e. for a given motion of the input link, all other links undergo unique definite motions, the mechanism is termed as a 'constrained mechanism'. It is clear that such mechanisms possess one degree of freedom. Degree of freedom denotes the number of independent quantities required to completely specify the configuration.

Figure 1.5 shows two mechanisms. In case of the mechanism shown in Fig. 1.5a, only θ is enough to specify the configuration of the whole mechanism, whereas for the mechanism shown in Fig. 1.5b, two angles 'θ' and 'φ' (or any two independent quantities viz h_1 and h_2) are needed to completely specify the configuration of the mechanism.

1.2.3 Mobility and Range of Motion

For successful operation, it is essential for a mechanism to possess appropriate and adequate motion capabilities. Any kind of assembly does not guarantee that the resulting agglomeration of the links will generate desired movements. Some systematic approach helps the designer to ensure suitable motion characteristics of assembled systems. The whole matter can be divided into the following issues:

At first suitable constrained motion of the mechanism has to be ensured. This can be done by ensuring appropriate degree of freedom.

If the objective of the mechanism is to drive a unit then it is essential to ensure that the links have the required degree of movability and motion ranges. Just possessing ability to move (when the required degrees of freedom exists) does not ensure success as the range of movements required may not be possible.

Ascertaining the required degree of freedom is essential at the very beginning when the basic kinematic chain is conceived. At this stage, dimensions of the links are not important. At this stage, it will be fine to keep the discussions confined to planar mechanisms only where all members are confined to move in two-dimensional planes parallel to one another. In two-dimensional motion, any link (i.e. a rigid body) possesses 3 degrees of freedom. Or, in other words, three independent quantities are required to completely specify its position and orientation. If there are n links all together, the system can have $3n$ degrees of freedom. But the kinematic pairs eliminate some of these freedoms. As mentioned earlier, a lower pair (prismatic or revolute) allows only one relative freedom of motion of a body connected to a fixed element. So, a lower pair removes 2 degrees of freedom. Similarly, a higher pair allows two relative degrees of freedom of motion, and such a pair removes 1 degree of freedom. Hence, a mechanism with n links (one being fixed to serve as the frame) connected by j lower pairs and h higher pairs possesses F degrees of freedom which is given by the following relation:

$$F = 3(n - 1) - 2j - h \tag{1.2.1}$$

It has been mentioned earlier that for a unique relationship between the input and output movements to exist, a mechanism must possess one degree of freedom, or $F = 1$. Using this in (1.2.1), the following relationship among n, j and h can be obtained:

$$1 = 3(n - 1) - 2j - h$$

or

$$3n - 2j - h = 4. \tag{1.2.2}$$

The above relation is called Grübler's criterion for constrained mechanisms. Of course there are special geometrical situations when the criterion fails, but for a perfectly general case, this relation must be satisfied. Figure 1.6a–d shows four constrained mechanisms.

Sometimes, just by observation, one may find out whether a mechanism is constrained or not. But for more complex cases, the Grübler's criterion may be very useful for routine checks. For situation where more than two links are connected by one hinge (or revolute pair), then the number of degrees of freedom eliminated by the hinge is given by

$$(i - 1)2$$

where i is the number of links connected at the hinge. Hence, if j_i be the number of hinges each one connecting i links, then the degree of freedom of a mechanism is given by

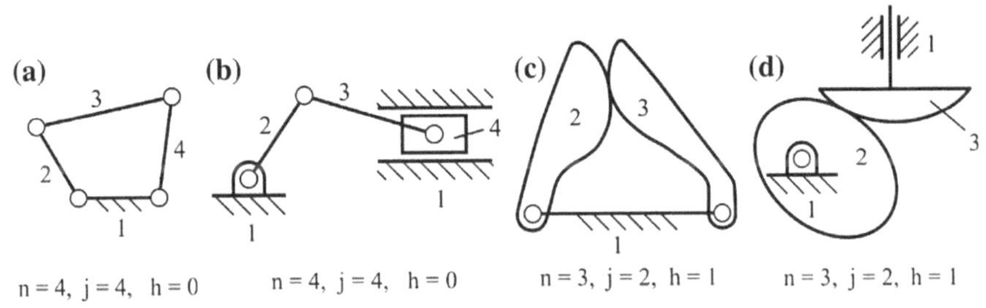

Fig. 1.6 Constrained mechanisms of different types. **a** Four-link mechanism with revolute joints only. **b** Four-link mechanism with revolute and prismatic pairs. **c** Three-link mechanism with one higher pair and two revolute pairs. **d** Three-link mechanism with one higher pair, one revolute and one prismatic pair

$$F = 3(n-1) - 2\sum_{i=2}^{m}(i-1)j_i - h \tag{1.2.3}$$

where m is the maximum number of links connected at a hinge and j_i is the number of hinges each connecting i links. To illustrate the usefulness of Grübler's criterion, the two cases shown in Fig. 1.7a, b may be taken up.

It is not easy to determine the number of degrees of freedom for the two mechanisms just by observation, but the use of Grübler's criterion provides the answers readily as demonstrated below.

For the mechanism shown in Fig. 1.7a $n = 11$, $j_2 = 9$, $j_3 = 1$, $j_4 = 1$, $h = 1$. This yields using (1.2.3)

$$\begin{aligned} F &= 3(11-1) - 2(2-1)9 - 2(3-1)1 - 2(4-1)1 - 1 \\ &= 30 - 18 - 4 - 6 - 1 \\ &= 1 \end{aligned}$$

So, the mechanism is a constrained one. For the other mechanism shown in Fig. 1.7b, $n = 12$, $j_2 = 10$, $j_3 = 1$, $j_4 = 1$, $h = 0$. Hence, one gets

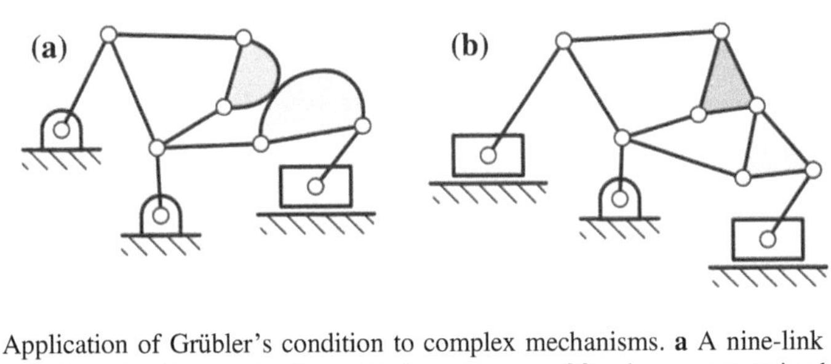

Fig. 1.7 Application of Grübler's condition to complex mechanisms. **a** A nine-link constrained mechanism. **b** A twelve-link mechanism with three degrees of freedom unconstrained mechanism

$$
\begin{aligned}
F &= 3(12-1) - 2(2-1)10 - 2(3-1)1 - 2(4-1)1 \\
&= 33 - 20 - 4 - 6 \\
&= 3
\end{aligned}
$$

So, the mechanism is not a constrained mechanism. Similar analysis and relationships exist for spatial mechanisms also; but, obviously, those are somewhat more complex.

Once the number of degrees of freedom is as desired, the mobility range needs to be determined in some cases. To illustrate the matter, let the case of a simple pin-jointed four-link mechanism be considered as shown in Fig. 1.8a. Depending on the links' relative dimensions, three different types of motion characteristics can be obtained.

In one case, the input link (say link 2) makes complete rotation, whereas the output link (say link 4) rocks to and fro as indicated in Fig. 1.8b. This type of mechanisms is called 'crank-rocker'. In another situation, both the input and the output links make complete rotations as shown in Fig. 1.8c and such mechanisms are termed as 'double-crank' mechanisms. In the third possible situation, both the input and output links can make rocking motion only as depicted in Fig. 1.8d, and naturally, these are called 'double-rockers'. All the three types can be obtained with different inversions of a kinematic chain satisfying the following condition:

$$l_{max} + l_{min} \leq l_1 + l_2 \tag{1.2.4}$$

where l_{max} and l_{min} are the lengths of the longest and shortest links, and l_1 and l_2 being the lengths of the other two links, respectively. This condition is known as

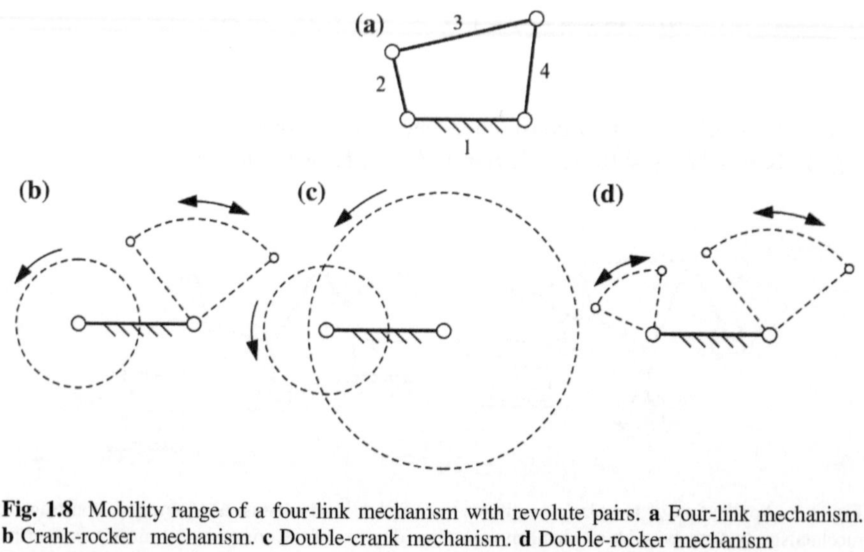

Fig. 1.8 Mobility range of a four-link mechanism with revolute pairs. **a** Four-link mechanism. **b** Crank-rocker mechanism. **c** Double-crank mechanism. **d** Double-rocker mechanism

Grashof's condition, and a linkage satisfying this condition is called a Grashof's linkage. If one of the links connected to the shortest link be grounded and the shortest link is used as a crank, one gets crank-rocker. If the shortest link is grounded, this inversion yields a double-crank. The chain results in a double-rocker when the shortest link is used as the coupler. When a chain does not satisfy condition (1.2.4), the chain is non-Grashofian and all inversions yield only double rockers.

However, one should note that for micromechanisms using flexure joints instead of revolute joints, the above study is not relevant as no link can make 360° relative rotation relative to the adjoining member as the whole structure is monolithic.

1.3 Kinematic Analysis and Synthesis of Plane Mechanisms

Once the basic topology features of a mechanism are finalized, it is important to decide the various dimensions so that mechanism fulfils its desired objectives and functions. Once the dimensions are worked out quite often an analysis of the mechanisms' displacement, velocity and accelerations is conducted to ensure proper functioning. Sometimes a second iteration of the synthesis (i.e. the step in which dimensions are decided) is carried out for refining the design. Quite often, the cyclic processes of synthesis and analysis are conducted for finalizing the design.

Since the initial phase of designing micromechanisms often follow the similar steps of conventional mechanism design. This section presents the rudimentary ideas of synthesis and analysis from kinematic point of view.

1.3.1 Types of Functions of Mechanisms

Mechanisms are used for widely varying types of functions as indicated below:

- Moving and manipulating an object
- Generating a path of desired shape
- Coordinating movements of links according to a prescribed function
- Generating a prescribed movement of the output link

In case of the first objective, a mechanism is used to transport and manipulate an object according to prescribed geometric characteristics. Quite often, gripping an object by the members of the mechanism is also a task. Figure 1.9 shows these types of functions in a few examples.

Path generation is an important task a mechanism has to perform. Figure 1.10a shows the generation of a prescribed path by a point P on the coupling link of a 4-link mechanism with 4 revolute pairs (4R mechanism).

By suitably choosing the dimensions of the links, a wide variety of paths can be generated. For path generation, there is a special class of mechanisms which are

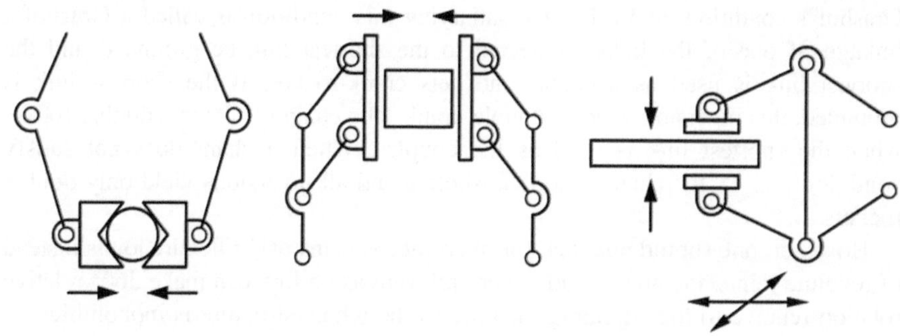

Fig. 1.9 Examples of gripping mechanisms

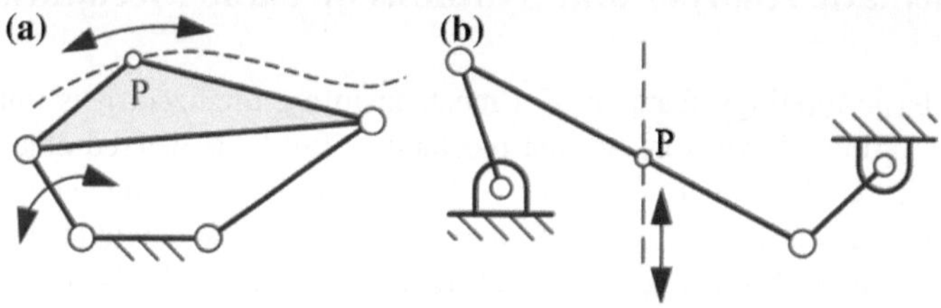

Fig. 1.10 Path-generating linkages. a Coupler-point curve generator. b A straight-line generator

used for generating a segment of a straight line using only revolute joints. These are called straight-line generators, and Fig. 1.10b shows an example that was first designed by James Watt. An important point needs to be noted here to recognize the importance of such mechanisms. It is already mentioned that revolute pairs in the form of simple hinge joints are the best and most reliable pairs for use in practical machines and mechanisms. From both reliability and cost of fabrication point of view, hinge joints are far superior to prismatic pairs, i.e. sliding joints. So, achieving the prescribed motion characteristics of a mechanism using hinge joints only requires appropriate synthesis. This is very relevant in case of micromechanisms as will be shown later.

Under certain circumstances, it becomes necessary to move a link maintaining a prescribed functional relationship with the input link's movement. These types of mechanisms are termed as function generators. Figure 1.11 shows the basic scheme for a particular case.

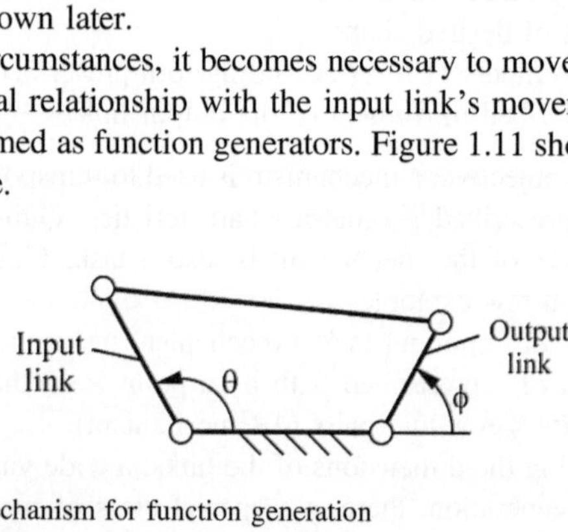

Fig. 1.11 Four-bar mechanism for function generation

If the rotation of the input link be given by the angle 'θ' and the output link rotation be devoted by ϕ, then for function generators, ϕ should be according to a prescribed function of angle θ. Thus,

$$\phi = f(\theta).$$

The simplest example of a function generator is a steering mechanism of a vehicle. This mechanism rotates the planes of the two front wheels in such a coordinated fashion so that the axes of rotation of the front wheels pass through the centre of rotation O while making a turn as shown in Fig. 1.12. To satisfy this condition, θ and ϕ must satisfy a functional relationship for the whole range of their movements.

There is another class of mechanisms which are used to generate specified reciprocating or oscillating motions.

Figure 1.13a, b shows two cases, usually called a crank-rocker and a slider-crank mechanism, respectively. In case of crank-rocker mechanisms, the design objective is to specify the rocking motion of the output link to be between positions 1 and 2, while the input crank makes continuous rotary motion. Similarly, the objective of design in case of slider-crank mechanisms is to specify the extreme positions of the slider for continuous rotation of the rotating crank.

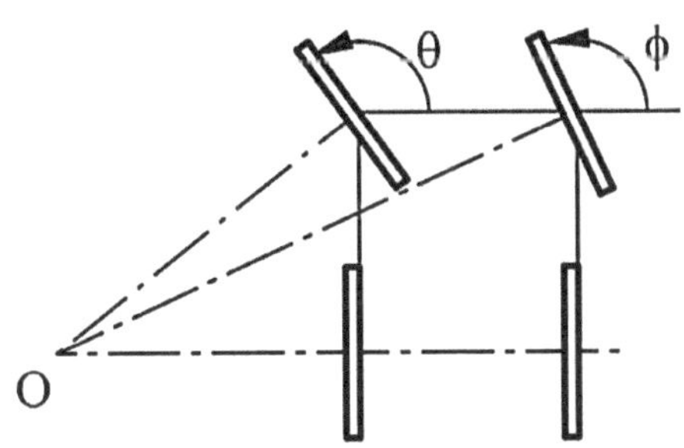

Fig. 1.12 Coordination of front wheel orientation while turning of a vehicle

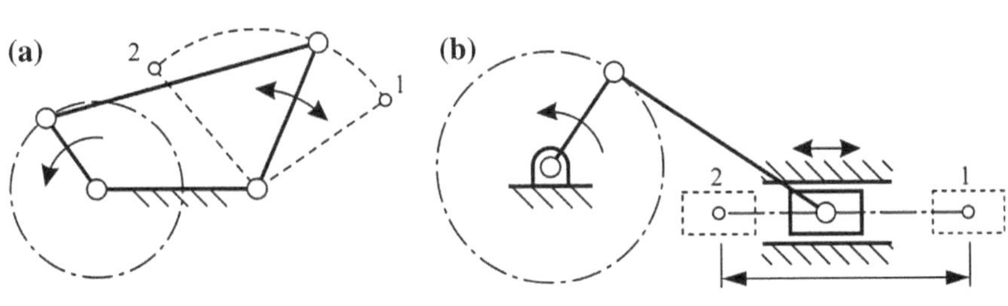

Fig. 1.13 Mechanisms for transforming motion. **a** Rotary-to-oscillatory transformation. **b** Rotary-to-reciprocating transformation

It must be remembered that the above types of requirements represent only the most commonly used mechanisms; there are many more types of mechanisms. In case of micromechanisms, however, still the majority of requirements can be covered by the above-mentioned cases.

1.3.2 Displacement, Velocity and Acceleration Analysis

Though the dimensions of a mechanism are necessary for any type of kinematic analysis and the initial step is synthesis, presenting the methods of analysis usually precedes a section on synthesis. Thus, it will be assumed that all the relevant dimensions of a mechanism have been determined by appropriate synthesis techniques. There are two approaches for kinematic analysis of mechanisms—graphical approach and analytical approach. The graphical approaches yield quick results, and it also helps to keep track of the physical reality. However, accuracy of the results obtained is not very high. Apart from this, when an analysis has to be repeated for a large number of times covering the full cycle of motion of a mechanism, graphical approach is not suitable. On the other hand analytical approaches require more work to begin with for developing the programme. But once that is done, the process can be repeated with ease as many times as one requires. The accuracy of the results obtained is also quite high. With the easy availability of computer these days, analytical approaches have become very attractive. Graphical approaches are still in use for obtaining an approximate solution to start the computational method for synthesizing mechanisms.

Displacement Analysis: In displacement analysis, the objective is to find out the positions of various links of a constrained mechanism corresponding to the various positions of the input link. The mechanism under consideration is drawn to a suitable scale. In case of constrained mechanisms, the positions of the all members can be determined once the position of the input link is given, by directly drawing the mechanism with the prescribed input link positions. The procedure can be explained with the help of the following example.

Figure 1.14 shows a six-link mechanism. As can be noted by observation, the mechanism has 6 links and 7 simple revolute joints. Hence, it satisfies Grübler's criterion, and the mechanism is a constrained one. Now to carry out displacement

Fig. 1.14 A six-link constrained mechanism

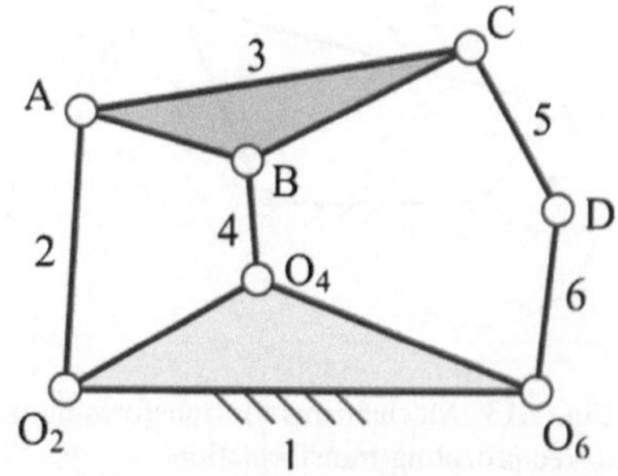

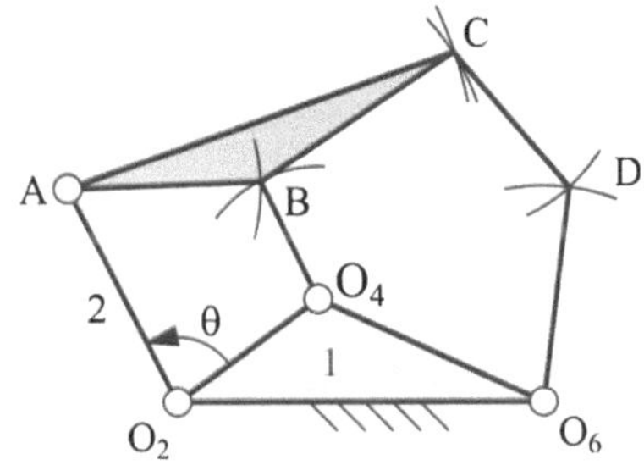

Fig. 1.15 Graphical displacement analysis of a six-link constrained mechanism

Fig. 1.16 Links and angles of a four-bar mechanism

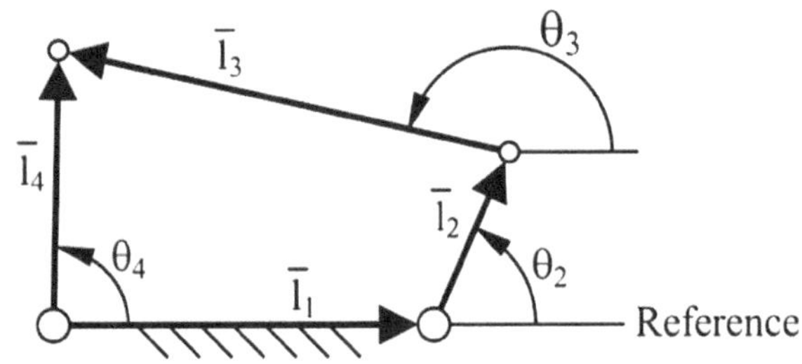

(or position) analysis, the mechanism needs to be drawn for given input link (say 2) positions. First, the ground link $O_2O_4O_6$ is drawn to scale. Next, the input link 2 is drawn for prescribed position specified by θ. Thus, the line O_2A is known. Next, an arc of radius equal to length AB is drawn, and with O_4 as the centre, another arc of radius equal to O_4B is drawn; these two arcs intersect at a point that specifies the position of point B as shown in Fig. 1.15.

With A and B as centres, two arcs with radii AC and BC, respectively, are drawn to intersect at C. Then, two arcs are drawn with C and O_6 as centres with radii equal to lengths CD and O_6D, respectively. The point of intersection specifies the position of D. From the figure, positions of all members are determined for a given value of θ and displacements of all other members can be found out.

Because of easy availability of computational facility, analytical approach is quite convenient. Apart from being easy for repetitive exercise and results being more accurate, analytical approach is applicable for situations when a graphical approach is not possible.[1] In case of simple mechanisms such as 4R or slider-crank mechanisms, direct analytical solutions are possible. The analysis for 4R mechanism is presented below.

Figure 1.16 shows a simple four-bar mechanism in which link 1 is fixed. The links are represented by the vectors $\bar{l}_1$, $\bar{l}_2$, $\bar{l}_3$, and $\bar{l}_4$, which constitute closed loop satisfying the condition

[1] It should be noted that in many cases involving complex mechanisms graphical approach cannot help as triangulation process fails.

$$\bar{l}_1 + \bar{l}_2 + \bar{l}_3 + \bar{l}_4 = 0 \tag{1.3.1}$$

The angles made by the four vectors with the reference direction are θ_1 (=0), θ_2, θ_3 and θ_4. The objective of this displacement analysis is to find out expressions for θ_3 and θ_4 when the input link angle θ_2 is given. The loop closure equation (1.3.1) can be also written using complex notation to represent the vectors in the following form:

$$l_1 + l_2 e^{i\theta_2} + l_3 e^{i\theta_3} - l_4 e^{i\theta_4} = 0 \tag{1.3.2}$$

as $\theta_1 = 0$. The above complex equation results in the following two equations by equating the real and imaginary parts to zero separately.

$$l_1 + l_2 \cos\theta_2 + l_3 \cos\theta_3 - l_4 \cos\theta_4 = 0 \tag{1.3.3a}$$

and

$$l_2 \sin\theta_2 + l_3 \sin\theta_3 - l_4 \sin\theta_4 = 0 \tag{1.3.3b}$$

The solution of the above two nonlinear equations in the two unknowns θ_3 and θ_4, when θ_2 and the link lengths are prescribed, results in the displacement analysis. It can be shown that there are two possible solutions as presented below:

$$\theta_4{}^{(1)} = 2\tan^{-1}\left\{\left(a + \sqrt{a^2 + b^2 - c^2}\right)\Big/(b + c)\right\}$$
$$\theta_3{}^{(1)} = 2\tan^{-1}\left\{\left(a + \sqrt{a^2 + b^2 - d^2}\right)\Big/(b + d)\right\}$$

and

$$\theta_4{}^{(2)} = 2\tan^{-1}\left\{\left(a - \sqrt{a^2 + b^2 - c^2}\right)\Big/(b + c)\right\}$$
$$\theta_3{}^{(2)} = 2\tan^{-1}\left\{\left(a - \sqrt{a^2 + b^2 - d^2}\right)\Big/(b + d)\right\}$$

where $a = \sin\theta_2$, $b = \cos\theta_2 + (l_1/l_2)$, $c = (l_1/l_2)\cos\theta_2 + \left\{\left(l_1^2 + l_2^2 + l_4^2 - l_3^2\right)/2l_2l_4\right\}$ and $d = -\left[(l_1/l_3)\cos\theta_2 + \left\{\left(l_1^2 + l_2^2 + l_3^2 - l_4^2\right)/2l_2l_3\right\}\right]$.

Because of the simplicity of the example, a closed-form solution of the nonlinear equations is possible. But in general, solution to the nonlinear equations resulting from the loop closure equations (more than one) is not possible and the problem has to be solved numerically with the help of computers.

Once the displacement analysis is accomplished and all the position parameters are determined, the next step is velocity and acceleration analyses. But the advantage with velocity and acceleration analysis is that while using analytical approach, the resulting equations are linear in terms of the unknown quantities, making the solution much *easier* than displacement analysis. Although traditionally graphical approaches for velocity and acceleration analyses are very popular, it is not being presented in details here and the reader can refer to any standard textbook on kinematic analysis. Only some important concepts are going to be presented here.

The most important characteristic of a rigid body (i.e. the link of a mechanism) is that the distance between two points on it remains fixed. In other words, the relative velocity between these two points must be along a direction perpendicular to the line joining these two points.

In other words, the components of the velocities $\bar{v}_A$ and $\bar{v}_B$ of two points A and B must have the same components along the line AB as shown in Fig. 1.17. Thus, once $\bar{v}_A$ is prescribed, $\bar{v}_B$ must be such that its tip must lie on the line X–X as indicated.

Another important concept in kinematic analysis is the concept of instantaneous centre of rotation. When a body's motion with respect to the fixed frame is given by specifying the instantaneous velocities of any two points A and B as shown in Fig. 1.18a, the body may be assumed to be making rotation instantaneously about an instantaneous centre I that is the point of intersection of the two normal to $\bar{v}_A$ and $\bar{v}_B$ drawn as points A and B. So the instantaneous angular velocity ω will be given by the following relation:

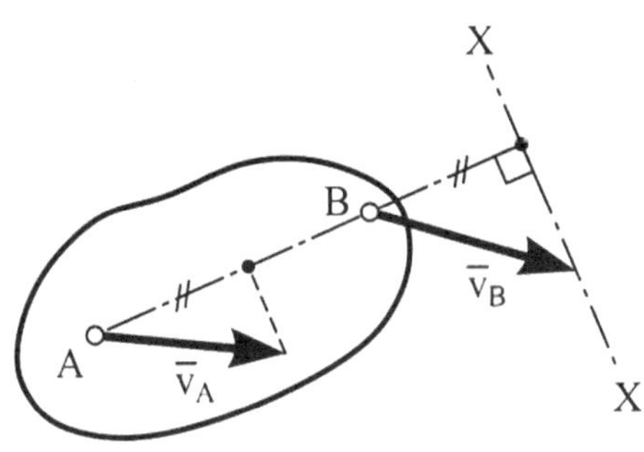

Fig. 1.17 Relative velocity between two points on a rigid body

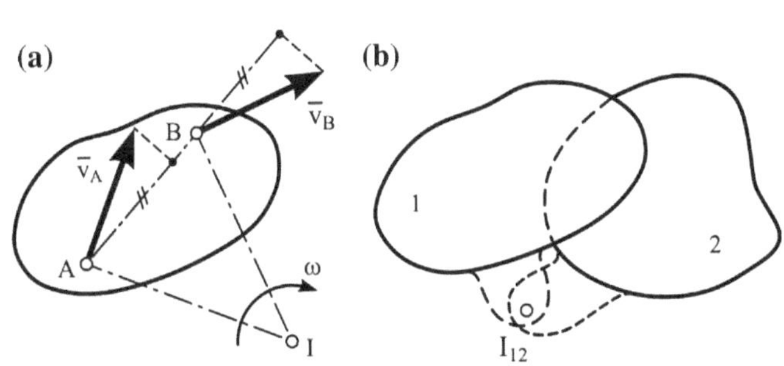

Fig. 1.18 Instantaneous centre of rotation of a rigid body. **a** Determination of instantaneous centre. **b** Relative instantaneous centre of two moving bodies

$$\omega = \frac{|\bar{v}_A|}{AI} = \frac{|\bar{v}_B|}{BI} \tag{1.3.4}$$

It is obvious that the instantaneous velocity of any other point C on the body can be readily found out. The magnitude of the velocity will be

$$|\bar{v}_C| = \omega \cdot CI$$

And the direction will be perpendicular to the line IC drawn at C. A related concept, relative instantaneous centre, is also very useful in kinematics. If the two bodies 1 and 2, shown in Fig. 1.18b, be in motion, then there exists a point I_{12} that is the relative instantaneous centre of the two moving bodies. The meaning is that a particle on body 2 located at the position I_{12} and another particle on body 2 located at I_{12} have same velocities at the instant. Or in other words, the particles have no relative motion, and instantaneously, the two bodies can be hinged at I_{12} without affecting the motion of the system at the instant. So, it is obvious that any hinge joint connecting two bodies represents the permanent relative instantaneous centre of the two bodies. A very useful theorem, named as Aronhold–Kennedy theorem, states that three relative instantaneous centres of three bodies in relative motion must lie along a straight line. The proof is simple and can be found in any standard textbook on kinematics. Using this quick velocity analysis can be achieved in many cases. The example of 4R mechanism is presented below.

For the mechanism shown in Fig. 1.19, the instantaneous angular speed of link 2 is prescribed as ω_2 and the task is to find out the instantaneous angular velocity ω_4 of link 4. First, it should be noted that points O_2, A, B and O_4 are the relative instantaneous centres I_{12}, I_{23}, I_{34} and I_{14}. From Aronhold–Kennedy's theorem, I_{12}, I_{14} and I_{24} must lie along a straight line that is found by joining O_4 and O_2. Similarly, I_{23}, I_{34} and I_{24} also must lie along a straight line found by joining A and B. Thus, I_{24} that lies on both of these lines is at the point of intersection of these two lines as shown in the figure. Now, two particles, one on link 2 and the other on link 4, located at the location I_{24}, have same velocities. This yields

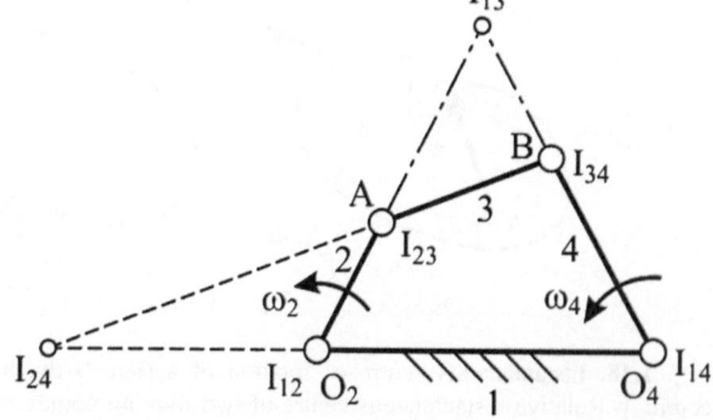

Fig. 1.19 Velocity analysis of a four-bar mechanism using instantaneous centre method

$$\omega_2 \cdot O_2I_{24} = \omega_4 \cdot O_4I_{24}$$

From this relation, the result is readily found as follows:

$$\omega_4 = \omega_2 \cdot \frac{O_2I_{24}}{O_4I_{24}}$$

Kinematic analysis using analytical approach is comparatively simpler as the equations to be solved are linear in terms of the unknown quantities. Once the displacement analysis is over, velocity analysis can be carried out in a straight-forward manner. Again, when the instantaneous velocities are determined, acceleration analysis can follow directly. Like in the case of displacement analysis, to begin with, the kinematically independent loops are identified and the loop closure equations are written in complex notation. To explain the procedure, the following example will be used. The system under consideration is the common quick return mechanism shown in Fig. 1.20a. The position parameters in this problem are θ_2 (input), θ_4 (output) and $|\bar{l}_4|$ (as the length of this vector is not constant).

These quantities are determined by displacement analysis with θ_2 prescribed. The vectorial loop closure equation is

$$\bar{l}_1 + \bar{l}_2 - \bar{l}_4 = 0 \tag{1.3.5}$$

And using complex notation for vectors (1.3.5) becomes

$$l_1 + l_2 e^{i\theta_2} - l_4 e^{i\theta_4} = 0 \tag{1.3.6}$$

as $\theta_1 = 0$ according to the chosen reference (Fig. 1.20b). Differentiating the LHS of (1.3.6) with respect to time

$$il_2\dot{\theta}_2 e^{i\theta_2} - il_4\dot{\theta}_4 e^{i\theta_4} - \dot{l}_4 e^{i\theta_4} = 0 \tag{1.3.7}$$

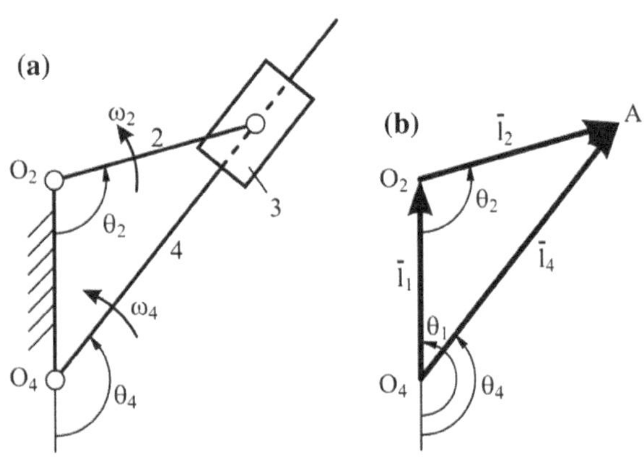

Fig. 1.20 Analytical approach of kinematic analysis of a four-link mechanism. **a** Parameters of a quick return mechanism. **b** Link vector loop

Equating the real and imaginary parts on the LHS separately equal to zero yields the following two linear equations with $\dot{\theta}_4$ and $\dot{l}_4$ as unknowns when $\dot{\theta}_2$ is prescribed:

$$l_4 \cos\theta_4 \cdot \dot{\theta}_4 + \sin\theta_4 \cdot \dot{l}_4 = l_2 \cos\theta_2 \cdot \dot{\theta}_2$$
$$l_4 \sin\theta_4 \cdot \dot{\theta}_4 - \cos\theta_4 \cdot \dot{l}_4 = l_2 \sin\theta_2 \cdot \dot{\theta}_2$$

Solving, one gets the analytical expressions for the unknown velocity parameters as follows:

$$\dot{\theta}_4 = l_2\dot{\theta}_2 \cos(\theta_2 - \theta_4) \Big/ l_4 \tag{1.3.8a}$$

and

$$\dot{l}_4 = l_2\dot{\theta}_2 \sin(\theta_4 - \theta_2) \tag{1.3.8b}$$

To carry out acceleration analysis (1.3.7) is differentiated with respect to time once more, and the following equation is obtained:

$$il_2\ddot{\theta}_2 e^{i\theta_2} - l_2\dot{\theta}_2^2 e^{i\theta_2} - il_4\ddot{\theta}_4 e^{i\theta_4} + l_4\dot{\theta}_4^2 e^{i\theta_4} - \ddot{l}_4 e^{i\theta_4} - i\dot{l}_4\dot{\theta}_4 e^{i\theta_4} = 0 \tag{1.3.9}$$

It should be noted that in the above equation, all quantities except $\ddot{\theta}_4$ and $\ddot{l}_4$ are known as $\ddot{\theta}_2$ is prescribed. Following a similar technique adopted in the case of velocity analysis, two linear equations with the two unknown acceleration terms are obtained solving which the following analytical solutions are found out:

$$\ddot{\theta}_4 = \frac{l_2\dot{\theta}_2^2 \sin(\theta_4 - \theta_2)}{l_4} - \frac{\dot{l}_4\dot{\theta}_4}{l_4} + \frac{l_2}{l_4}\cos(\theta_4 - \theta_2)\ddot{\theta}_2$$

and

$$\ddot{l}_4 = l_4\dot{\theta}_4^2 - l_2\dot{\theta}_2^2 \cos(\theta_4 - \theta_2) + \ddot{l}_2 \sin(\theta_4 - \theta_2) \cdot \ddot{\theta}_2$$

For more complex systems, more than one loop equations need to be formed. For large systems, computer solutions become convenient.

Kinematic analysis becomes necessary after a mechanism is synthesized. The objective is to primarily check whether the synthesized mechanisms behave and perform as per the prescribed requirements. In the next subsection, the basic methodologies for kinematic synthesis are presented.

1.3.3 Kinematic Synthesis of Plane Mechanisms

Synthesis of a mechanism starts with basic number synthesis in which it is decided how many links should be used and the basic topological characteristics of the *kinematic* chain. In number synthesis, it is ensured that the designed mechanism enjoys the appropriate number of degrees of freedom and the correct range of mobility. Generally, more number of links are used when more complex functions have to be generated and/or higher level of accuracy in the performance parameters is essential. Otherwise, a mechanism should be chosen with minimum possible number of moving members. This improves the reliability and keeps the cost low. In many occasions, this step of synthesis also depends on the experience and intuitive judgement of designers.

In general, micromechanisms are composed of minute-sized components as their macroscopic components. However, the more recent trend is to design monolithic structures with localized compliances for micromechanisms. The matter is explained in Fig. 1.21 that as fabrication and assembly of hinged members are extremely difficult for microsized objects, the hinges are replaced by deep notches/grooves at the equivalent locations.

These localized compliances provide limited relative rotational movement to the adjoining rigid links as indicated in Fig. 1.21b. Though such flexure joints cannot provide full rotational degree of freedom like hinges, they play the role of equivalent hinge joints with somewhat limited rotational freedom. Thus, within a limited range of relative motion, the system shown in Fig. 1.21b behaves like a 4R mechanism shown in Fig. 1.21a. It is needless to mention that fabrication of such monolithic units with flexure hinges is much simpler than the fabrication of multicomponent micromechanisms. There are two major approaches for designing micromechanisms—one approach is called continuum synthesis approach and the other is kinematic synthesis approach. Both of them will be discussed in detail later, but it will be enough to point out at this stage that in kinematic synthesis approach first a conventional mechanism is synthesized to generate the required movements. Subsequently, the kinematic pairs are replaced by suitable flexure hinges. Thus, it is obvious that in the kinematic synthesis approach one should use only revolute pairs (i.e. hinges) only.

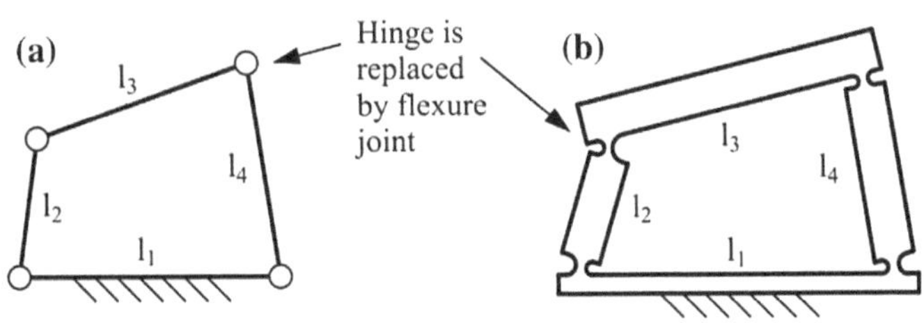

Fig. 1.21 Replacement of hinges by flexure joints. **a** Four-bar mechanism with four hinge joints. **b** Kinematically equivalent monolithic compliant mechanism

Number Synthesis: As mentioned earlier, a mechanism must have appropriate degrees of freedom for proper functioning. If there is only one input motion, then the mechanism must possess a single degree of freedom so that a unique output motion is generated. When two independent input motions are involved, the mechanism should possess two degrees of freedom to provide proper functioning of movements. The number of degrees of freedom of a plane mechanism can be expressed as follows:

$$F = 3(n - 1) - 2(j_1 + 2j_2 + 3j_3 + \cdots) \tag{1.3.10}$$

where j_i is the number of hinges joining $(i + 1)$ number of links. It should be also noted that there are no hinge pairs involved for reasons easily comprehended. F and n represent the number of degrees of freedom and the total number of links (including the fixed frame), as before. A kinematic chain should be also examined to avoid formation of structural truss members (when three links are hinged with one another). So a six-link mechanism should be either as shown in Fig. 1.22a, b but not as shown in Fig. 1.22c.

Once the topological structure of the kinematic chain representing a mechanism is finalized, the next step is the determination of the kinematically relevant dimensions. In hinged mechanisms, these dimensions are the distances among the hinges on every individual link.

Dimensional Synthesis: As in the case of kinematic analysis in dimensional synthesis also there exist both graphical and analytical approaches. Graphical approach, as usual, leads to quick results. But the accuracy of the results may not be high, and graphical approaches may not be very suitable when the process needs repetition for a large number of times. Of course, quite often, an approximate result from graphical method is utilized as the starting point for analytical approach. Besides, synthesis by graphical approach keeps track of the physical reality and, in some situations, helps the designers to discard physically meaningless solutions. Synthesis can be again exact or approximate. In case of exact synthesis, the prescribed performance and the input–output relation exactly satisfied over the whole range of operation. But in most cases, the desired task cannot be executed in an exact manner and an approximate synthesis is adopted. There are primarily two approaches for approximate synthesis. These are (i) precision (or accuracy) point approach and (ii) error minimization approach. The major types of mechanism

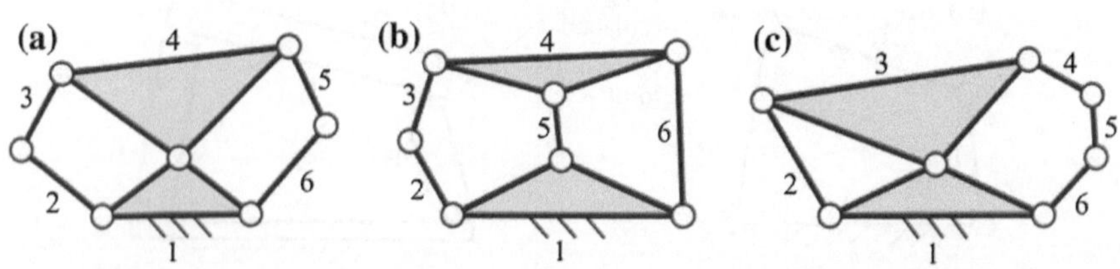

Fig. 1.22 Different kinematic structures of six-link mechanism. **a** Watt's mechanism. **b** Stephansion's mechanism. **c** Six-link mechanism with unacceptable kinematic structure

synthesis problems are for (i) motion generation, (ii) path generation and (iii) function generation as mentioned earlier.

To start with, the 'precision (or accuracy) point approach' will be discussed. In this approach, the required conditions are exactly satisfied only at a limited number of positions in the whole range of movements. These positions are called accuracy points and spaced in a manner to keep the maximum error in the whole range limited to a minimum. These optimally spaced points are called 'Chebyshev's accuracy points'. A mechanism is synthesized based on this finite number of precision points. Graphical methods use this approach.

In the error minimization approach, the error is minimized in the whole range in an overall sense. The approach is analytical in nature, and this approach is particularly suitable for mechanisms with more than four links.

Accuracy point method yields better result if the number of accuracy points is increased and optimally spaced. However, there is a limit to the number of accuracy points.

Graphical procedure for motion generation using 4R linkage is demonstrated below with gradually increasing number of accuracy points.

Figure 1.23 shows the basic approach for moving a rigid body for satisfying two positions. The body has to take the two positions marked ① and ②, and a 4R linkage is to be designed to ensure that.

Two suitable points on the body A and B are chosen first. For position ①, the location of these points is A_1 and B_1. These points go to locations A_2 and B_2 when the body takes its second stipulated position ②. A_1 and A_2 are joined, and the perpendicular bisector I is drawn. Similarly, B_1 and B_2 are joined, and the perpendicular bisector II is constructed. Two suitable positions on these lines I and II, O_2 and O_4, respectively, are chosen. $O_2A_1B_1O_4$ is the 4R mechanism corresponding

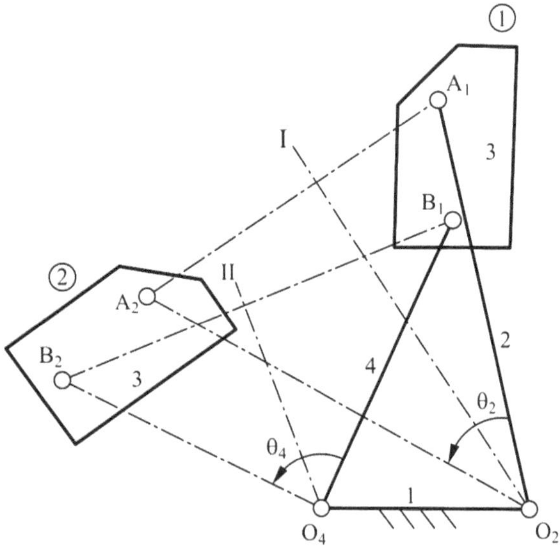

Fig. 1.23 Motion generation of a rigid body with two prescribed positions

to the first location. This mechanism also ensures that the body takes the second desired position with the mechanism being $O_2A_2B_2O_4$ configuration. As A and B can be any two points in a plane, the number of possibilities is equal to ∞^4; again, as O_2 and O_4 can be any two points on the lines I and II, the number of possibilities becomes ∞^2, making the total number of possibilities for the mechanism equal to ∞^6.

If the body has to take three prescribed positions as shown in Fig. 1.24, the fixed hinges O_2 and O_4 are suitably chosen as any two points as shown in Fig. 1.24a. Next, the locations of O_2 and O_4 relative to the body in its positions ② and ③ are superimposed on the figure for the first location as $O_2^{(2)}$, $O_4^{(2)}$, $O_2^{(3)}$ and $O_4^{(3)}$ as shown in Fig. 1.24b. (This is often called the inverted position.) The point of intersection of two perpendicular bisectors, I and II, of A_1 represents the centre of circle that passes through $O_2^{(1)}$, $O_2^{(2)}$ and $O_2^{(3)}$. Hence, it can be connected to the fixed hinge O_2 by a link O_2B_1. Similarly, the location of B_1 is found, and $O_2A_1B_1O_4$ represents the desired 4R mechanism. As O_2 and O_4 can be chosen freely in the plane, the number of possibilities is ∞^4.

With more than three positions as prescribed, the problem becomes complicated as a circle cannot be drawn passing through any four points. When four positions ①, ②, ③, and ④ are prescribed as shown in Fig. 1.25a, there exists a curve a point on which lies on a circle. This curve is called the 'circle-point' curve. Similarly, the locus of the corresponding centres of the circles is called the 'centre-point' curve. The circle-point and the centre-point curves for the four specified positions are shown in Fig. 1.25. The moving hinges on the body can be chosen on the circle-point curve at any two suitable locations at A and B as indicated in Fig. 1.25b.

Once A and B are chosen, the corresponding centres O_2 and O_4 are fixed on the centre-point curve without any choice. The 4R mechanism that can take the rigid body through the four specified positions is O_2ABO_4. The total number of possibilities is ∞^2.

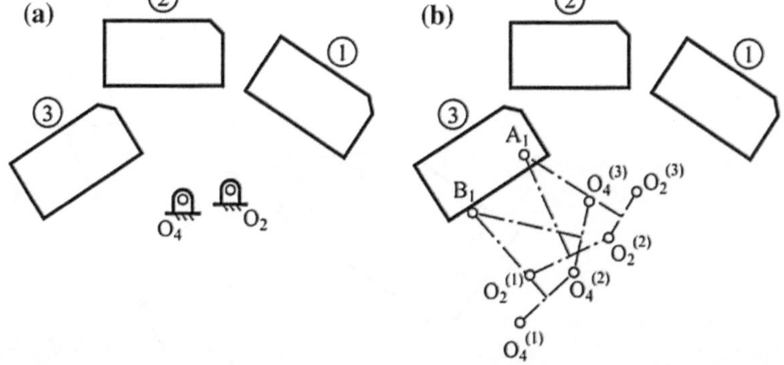

Fig. 1.24 Motion generation with three prescribed positions. **a** Three successive positions of the body. **b** Determination of fixed hinges with prescribed moving hinges

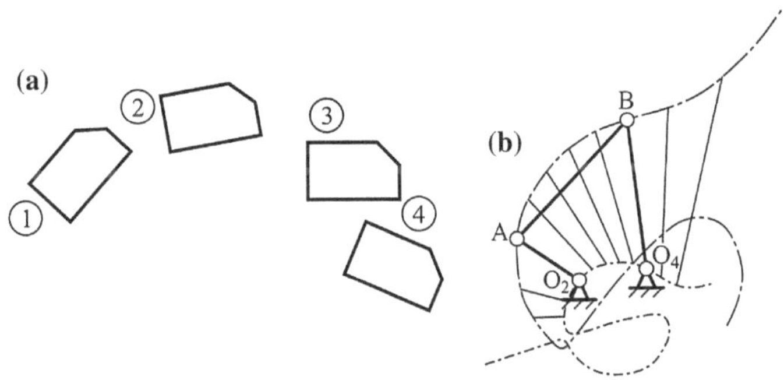

Fig. 1.25 Motion generation for four positions. **a** Prescribed four positions. **b** Circle-point and centre-point curves

If a body has to be guided through five specified positions, then there exist only a few points which move along circles and the number of possibilities is ∞^0. These points are called *Burmester* points. It is now clear that if a body has to be guided through 6 or more number of positions, the number of possibilities is ∞^{-2}, i.e. nil.

Sometimes, it is desired to generate a prescribed curve (or path) by a point on a link of a mechanism. The graphical procedure using three accuracy points is described below.

Let the path be as shown in Fig. 1.26 on which three accuracy points C_1, C_2, and C_3 are chosen suitably. A 4R mechanism has to be synthesized so that a point of the coupling link (as that is the only member in a 4R mechanism that does not make pure rotation about a fixed hinge) occupies these three positions in succession. To start with, corresponding to the first position of the mechanism, hinge A_1 is chosen at a suitable point after the location of the two fixed hinges, O_2 and O_4 are selected. So, a part of the mechanism becomes $O_2A_1C_1$ where C_1 is the first position of the coupler point. The task ahead is to find out the other moving hinge B_1, in its first location. As hinge B_1 has to be connected to the fixed hinge O_4, it must lie at the point that remains equidistant from O_4. This can be done by finding the inverted positions of O_4 when the coupler point of the mechanism, C, occupies the second and third accuracy points C_2 and C_3 (Fig. 1.27).

Fig. 1.26 Path generation by a coupler point

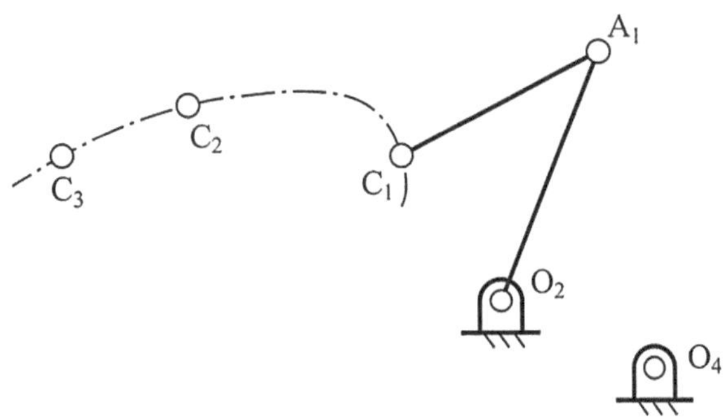

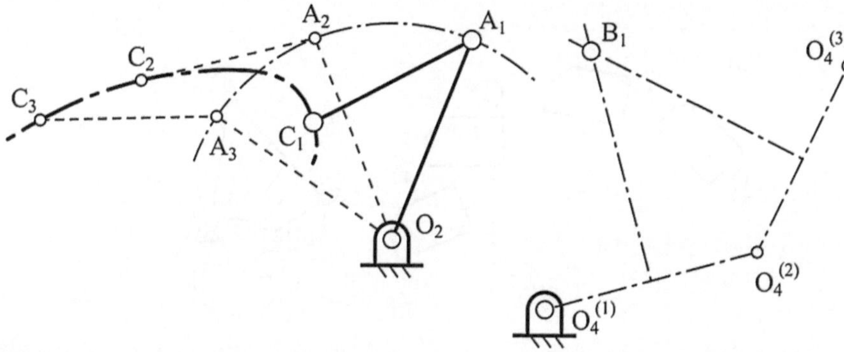

Fig. 1.27 Graphical synthesis procedure for path generation

Fig. 1.28 Synthesized final path-generating mechanism

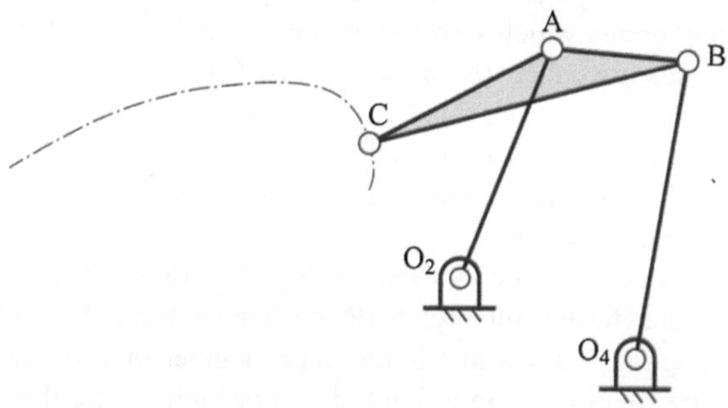

The inverted positions of O_4 (when superimposed on to the first configuration of the linkage) are $O_4{}^{(2)}$ and $O_4{}^{(3)}$. These are found out by triangulation using the distances C_2O_4 and A_2O_4 for $O_4{}^{(2)}$ and C_3O_4 and A_3O_4 for $O_4{}^{(3)}$. Next, the centre of the circle passing through $O_4{}^{(1)}$, $O_4{}^{(2)}$ and $O_4{}^{(3)}$ is found out, which is the desired location of B_1. The 4R mechanism is O_2ABO_4, and C is the coupler point (Fig. 1.28).

A very special type of path-generating mechanisms describe straight lines, which are important for many applications. Some straight-line mechanisms generate exact straight lines, whereas a majority describe approximate straight lines. To synthesize an approximate straight-line mechanism, the above procedure can be followed, while the three accuracy points occupy three suitably spaced positions on the straight line to be generated.

Figure 1.29 presents an example showing the synthesis of a straight-line mechanism. Suitable choice of the accuracy point locations is extremely important for good performance. Chebysev's accuracy points are the most commonly used choice. If the position parameter be denoted by x and x_i and x_f be its initial and final values and the number of accuracy points be k, then the location of the Chebysev's accuracy points is given by the following relation:

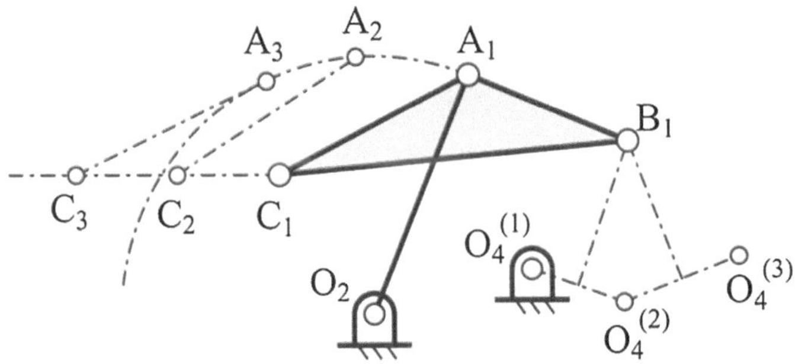

Fig. 1.29 Synthesizing straight-line path generator

$$x_l = \frac{x_i + x_f}{2} + \frac{x_f - x_i}{2}\cos\frac{(2l-1)\pi}{k}; \quad l = 1, 2, \ldots, k \tag{1.3.11}$$

Synthesizing function generators can be represented by a problem that can assure coordinated positions of the input and output links. For example, Fig. 1.30 shows the input link's three positions and the corresponding movements (i.e. rotations) of the output link.

Thus, two pairs of coordinated rotations of the input and the output links are specified. The graphical approach is again based upon the determination of the location of B, the other moving hinge, after O_2, O_4 and the input link O_2A's three positions are marked. This time inversions are based upon fixing the follower link at its first position. As the output link rotates through an angle $\theta_4{}^{12}$, relative to this link, the rest of the system rotates through $-\theta_4{}^{12}$. Similarly, when the follower rotates through $\theta_4{}^{13}$, the rest rotates through $-\theta_4{}^{13}$. Since the angle link O_2A makes with respect to the base link (O_2O_4), the prescribed location of A with respect to O_4O_2 can be found out. Thus, keeping the follower link fixed at its first position, the positions of A with respect to the follower at three configurations of the mechanism can be found out as shown in Fig. 1.31.

Once the three inverted positions $A_1{}^{(1)}$, $A_2{}^{(1)}$ and $A_3{}^{(1)}$ are found out, B_1 is found out as the centre of the circle passing through $A_1{}^{(1)}$, $A_2{}^{(1)}$ and $A_3{}^{(1)}$ as

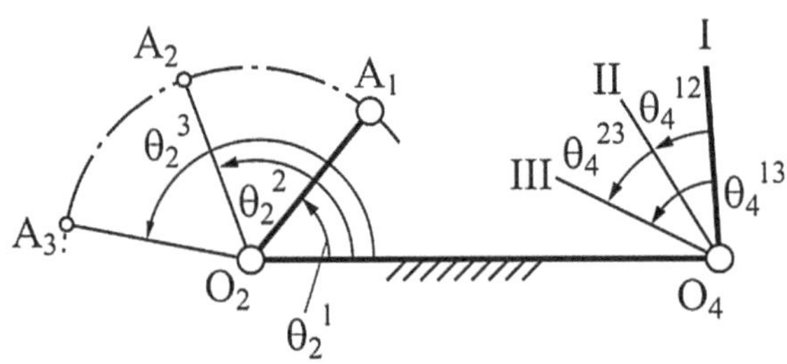

Fig. 1.30 Coordinated rotation of crank and follower links

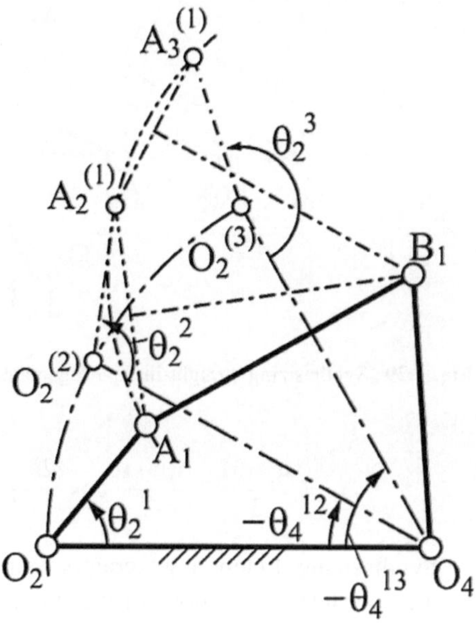

Fig. 1.31 Synthesis of function generator

indicated in the figure. Then, the 4R mechanism is given by $O_2A_1B_1O_4$ in its first position.

For designing micromechanisms using kinematic synthesis approach, first the hypothetical hinged linkage is designed. The above procedures can be used for the purpose for less accurate tasks.

Analytical Approach: The scope of graphical approach is somewhat limited as only approximate results can be achieved. Furthermore, more complex mechanisms may not be handled graphically in all cases. With the easy availability of computers, analytical approaches have gained popularity. Three-position synthesis of 4R mechanism using analytical approach is of course simple as shown below:

The input link makes angles θ_2^1, θ_2^2 and θ_2^3, whereas correspondingly the follower link makes angles θ_4^1, θ_4^2 and θ_4^3 with the fixed link as shown in Fig. 1.32a. Since only angular movements are concerned, only the three-link length

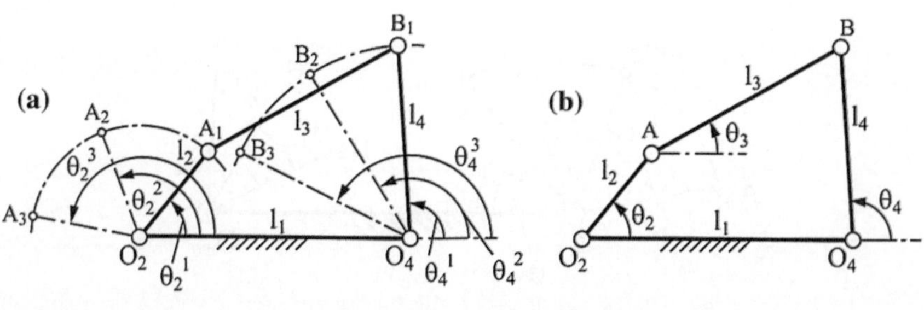

Fig. 1.32 Analytical procedure for synthesizing four-bar function generator

ratios (l_2/l_1), (l_3/l_1) and (l_4/l_1) are relevant and should be determined. From Fig. 1.32b,

$$l_3^2 = (l_1 + l_4\cos\theta_4 - l_2\cos\theta_2)^2 + (l_4\sin\theta_4 - l_2\sin\theta_2)^2$$

Expanding and then rearranging the terms, the above relation becomes as follows:

$$\begin{aligned}\cos(\theta_2 - \theta_4) &= \frac{l_1}{l_2}\cos\theta_4 - \frac{l_1}{l_4}\cos\theta_2 + \frac{l_1^2 + l_2^2 - l_3^2 + l_4^2}{2l_2l_4} \\ &= \lambda_1\cos\theta_4 - \lambda_2\cos\theta_2 + \lambda_3\end{aligned} \quad (1.3.12)$$

Now, the synthesized mechanism satisfies the above equation exactly at the specified three positions given by ${\theta_2}^1$, ${\theta_2}^2$ and ${\theta_2}^3$ with the corresponding values of ${\theta_4}^1$, ${\theta_4}^2$ and ${\theta_4}^3$. Substituting the values in (1.3.12), three linear simultaneous equations in three unknowns λ_1, λ_2 and λ_3 are obtained. Solving the equations, λ_1, λ_2 and λ_3 are found out which yield the values of the three link-length ratios l_1/l_2, l_1/l_3 and l_1/l_4. It is clearly seen that for a 4R mechanism, only three link-length ratios exist and only three coordinated positions can be specified. If four positions are specified, then besides the three-link length ratios, one more quantity needs to be kept as an unknown. If ${\theta_4}^1$ is kept as unknown and the three successive rotations of the follower are prescribed, then (1.3.12) can be expressed as follows:

$$\begin{aligned}\cos({\theta_2}^1 - {\theta_4}^1) &= \lambda_1\cos{\theta_4}^1 - \lambda_2\cos{\theta_2}^1 + \lambda_3 \\ \cos({\theta_2}^2 - {\theta_4}^1 - \Delta{\theta_4}^{12}) &= \lambda_1\cos({\theta_4}^1 + \Delta{\theta_4}^{12}) - \lambda_2\cos{\theta_2}^2 + \lambda_3 \\ \cos({\theta_2}^3 - {\theta_4}^1 - \Delta{\theta_4}^{13}) &= \lambda_1\cos({\theta_4}^1 + \Delta{\theta_4}^{13}) - \lambda_2\cos{\theta_2}^3 + \lambda_3 \\ \cos({\theta_2}^4 - {\theta_4}^1 - \Delta{\theta_4}^{14}) &= \lambda_1\cos({\theta_4}^1 + \Delta{\theta_4}^{14}) - \lambda_2\cos{\theta_2}^4 + \lambda_3\end{aligned}$$

where $\Delta{\theta_4}^{12}$, $\Delta{\theta_4}^{13}$, and $\Delta{\theta_4}^{14}$ are the prescribed rotations of the follower link from the still unknown first position. Thus, the above set of equations contain four unknowns, λ_1, λ_2, λ_3 and ${\theta_4}^1$. Solution of the above set of nonlinear equations is not easy, but computers can produce the result.

As mentioned earlier, it is always desirable to carry out kinematic analysis of a mechanism after it is synthesized. If found necessary, the synthesis process has to be repeated with altered initial choices to satisfy the prescribed requirements.

1.4 Actuators

As stated earlier, relative motion among the members is an essential feature of a mechanism to accomplish the task. The devices that provide the input motions to mechanisms are called actuators. Scientifically defining, an actuator is a device that transforms electrical, thermal or fluid flow into controllable mechanical motion. In the macroscopic level, the commonly used actuators are the following:

- Electric motors
- Hydraulic/pneumatic cylinder–piston drives
- Bimetallic thermostats
- Magnetostrictive devices
- Shape memory alloy-based drives
- Conducting smart polymers
- Piezoelectric devices

Electromagnetic energy is the most common source of energy for motors, and such motors are based upon the attractive and repulsive forces generated on conductors carrying current in a magnetic field. Motors and electric solenoids are the most common examples. As electromagnetic fields can change very fast, the response of these actuators is also very quick. Unfortunately, electromagnetic actuators are not that advantageous for downward scalability into the micro- and nanometre scales. Besides the required geometrical alignments between the directions of current flow, the magnetic field and the generated force make the design of planar systems difficult.

Electrostatic actuation is based upon the Coulomb force as the motive force. Electrostatic actuators are more favourably disposed towards downward scaling. These actuators are also fast in response and are also very energy efficient. Electrostatic motors with 100 μm diameter operating at 25–35 V at 2,500–15,000 rpm generating around 10–12 pN-m torque have been designed and fabricated at the early stage of microsystem technology.

Thermomechanical systems use the expansion of an element when heated as the main source of mechanical movement. Obviously, the range of motion in such actuators is very limited, but very large forces can be developed. To amplify the mechanical movements, bimetallic strips are used in which the coefficients of thermal expansion of the two metals are different. But with a configuration shown in Fig. 1.33, a substantial deflection δ can be achieved. If the lengths of the two strips differ by a small amount Δ after the temperature has raised, the radius of curvature will be given by $R \approx lh/\Delta$, where h is the distance between the neutral

Fig. 1.33 Basic principle of a bimorph

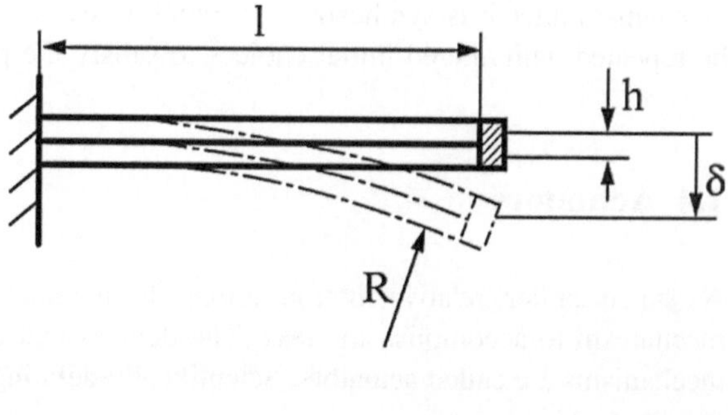

axes of bending of the two strips, l being the length of the strips. The lateral deflection δ can be found out as follows:

$$\delta \approx \frac{l\Delta}{h}$$

Thus, even if Δ is small, making h also very small, significant deflection can be obtained. As thermal actuators depend on the rate of heating and cooling in macroscopic scale, the device is not fast. But in microscale, the response can be very quick.

Piezoelectric crystals undergo small changes in dimensions when electrically charged at the two opposing faces. Typical materials are quartz (SiO_2), lead zirconate (PZT), polyvinylidene fluoride (PVDF), etc. Actuators using piezoelectric crystals possess quick response and good repeatability. These devices can generate large forces also. Current common usage includes its use in surface tunnelling electron microscopes for rapidly moving the sensing tip over a very small range. Piezoelectric actuators are also commonly used in ultrasonic transducers.

Smart materials possess inherent capacity to undergo large deformation on receiving the right stimulus. Shape memory alloys (SMA) occupy a significant place in the group of smart materials used in actuators. The shape memory effect takes place when a member, made of an SMA, is cycled above or below a specified transition temperature. The shape change is produced because of the alloy's transition from martensite to austenite phase as indicated in Fig. 1.34. At higher temperatures, the crystals attaining austenite phase shrink, causing the member to reach its original shape.

Another smart material in common use is the ionic polymer–metal composites (IPMC) or, in general, the electroactive polymers. These composites consist of a thin film with electrodes deposited on two faces. On application of a potential, the mobile cations move towards the negative electrode causing the film to stretch along the negative electrode causing it to bend as shown in Fig. 1.35. Of late, these types of electroactive polymers are also used for actuating devices. These films require an aqueous environment to keep the metallic cations mobile. Force developed is not large.

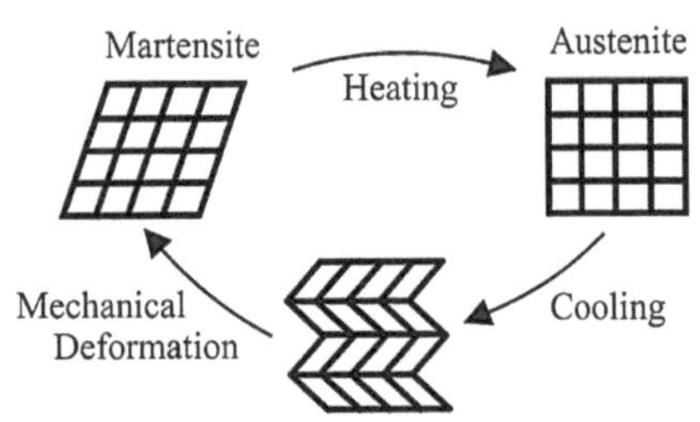

Fig. 1.34 Different stages of a shape memory alloy

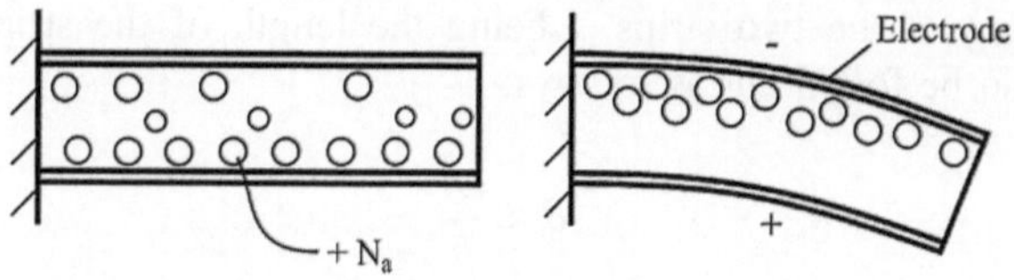

Fig. 1.35 Principles of ionic polymer–metal composite-based actuators

Some other types of exotic actuation systems are also in use. However, a direct downscaling of all the above types of actuators along with the micromechanisms may not be possible in many cases. The primary reason for this is the role of the scaling laws controlling various aspects of an actuator system. The detailed discussion on the scaling laws will be presented in a later section. Microactuators will also be presented in detail later in this volume.

Chapter 2
Miniaturization and Microsystems

Abstract Current trends of miniaturization and its need are presented. The impact of miniaturization on design principles for micromechanisms is discussed. The important role microsystems and miniaturization will play in the impending 3rd industrial revolution is also discussed in brief.

Keywords Miniaturization · Third industrial revolution · Microsystems

2.1 Miniaturization

To appreciate the subject of this volume 'Micromechanisms and Microactuators', it is desirable to first recognize the importance of the topic miniaturization as per the current trend of technological development. The matter of miniaturization is, in fact, closely related to the impending paradigm shift slowly taking place. A paradigm shift in technology brings overwhelming transformations in the society affecting almost all spheres of activities. In the recorded history, only twice such changes took place. These are usually termed as 'Industrial Revolution'.

The first industrial revolution took place during the period 1775–1830 when steam engines appeared in the scene as the first prime movers. Once steam power took over manufacturing became an important activity controlling, the world economy and technology became the most important factor in deciding world power. To recognize the phenomenon in a quantitative manner, the yearly pig iron production was taken as the estimating parameter. Figure 2.1 shows the growth of pig iron production, and the sudden increase in the rate is quite visible.

The second industrial revolution started in the 1960s and reached its maturity towards the turn of the millennium. Emergence of microelectronics, computers, communication systems, knowledge-based industry and the information technology transformed the society which we are lucky to experience personally. All these became major factors in the world economy. If one takes up the GDP of the western European countries and plots it over the years, one gets a figure as shown in

A. Ghosh and B. Corves, *Introduction to Micromechanisms and Microactuators*, Mechanisms and Machine Science 28, DOI 10.1007/978-81-322-2144-9_2

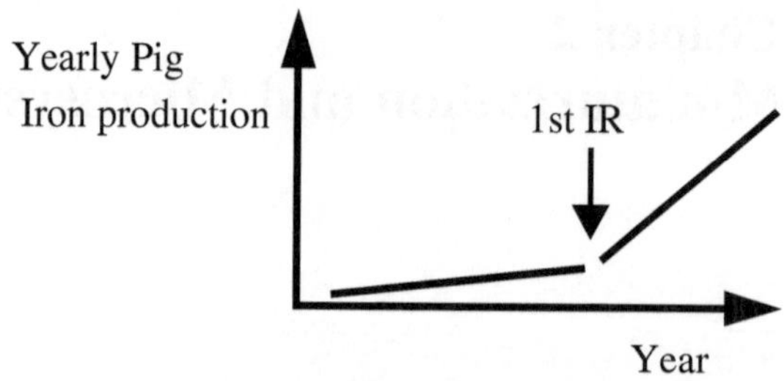

Fig. 2.1 Change in iron production rate due to onset of first industrial revolution

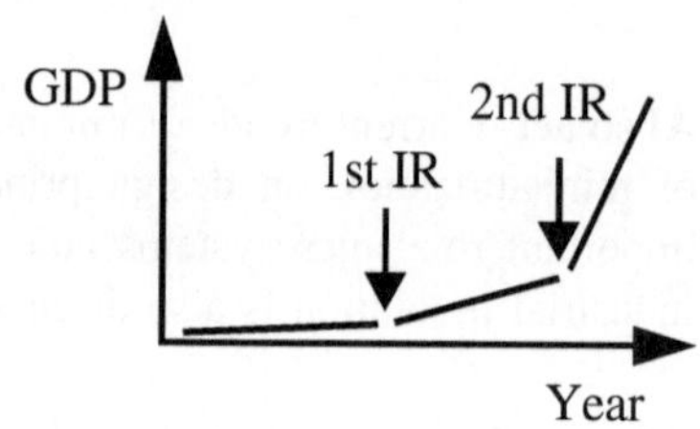

Fig. 2.2 Onset of the first and second industrial revolutions as indicated by the GDP growth

Fig. 2.2. The sudden changes in the slopes clearly indicate the revolutionary changes and paradigm shifts. The onset of the 2nd industrial revolution is very clearly visible.

It is well known that the revolutionary developments of the computer-based system took place primarily through a process of miniaturization of the electronics involved. Because of this, not only the size, but more importantly the cost has crashed down to every ones reach. However, in the 2nd industrial revolution, miniaturization was restricted to only two-dimensional electronic circuits, the primary function of those being manipulation of information. But it is slowly being recognized that three-dimensional miniaturization of devices and systems which physically interact with objects in the surrounding has far more potential to bring revolutionary changes in our society. It is predicted by many that economically viable technology to minimize 3D objects will form the basic foundation for the forthcoming 3rd industrial revolution.

2.1.1 Current Trend

Because of the progress of miniaturization technology, many devices are now produced in very small sizes. In fact, the progress in microelectronics technology has given birth to a new field—microelectromechanical systems (MEMS). As the VLSI chips are becoming more advanced, the fabrication facilities for old chips are becoming obsolete. Instead of dismantling the old set-ups, these are being effectively used in fabricating MEMS devices which require somewhat lesser degree of miniaturization than that for electronic chips. These readily available fabricational facilities have given excellent boost to the development of MEMS technology.

The current trend of technological development arises out of two major requirements. The modern and advanced systems are being provided with greater degree of intelligence and autonomy, and the other trend is towards making the physical devices smaller in size. In some cases, both objectives are being combined for producing miniaturized autonomous devices. Figure 2.3 depicts these trends symbolically. The technology for material manipulation is also advancing continuously, and smaller levels for material manipulation are being achieved.

Figure 2.4 shows how with time the capability to manipulate material is reaching lower and lower levels. Currently, technologists have acquired the capability to operate at nanometre levels and that is why 'nanotechnology' is drawing so much attention. It may not be a long time when the technologists will be in a position to control and manipulate individual atoms at least the trend is very clear.

As mentioned in the previous paragraph, one of the objectives of development is to make the futuristic systems more intelligent and autonomous. To a large extent the degree of intelligence with which a device can function depends on the availability of information. This will increase the requirements of sensing ability to many orders of magnitude more than the present systems. Thus, very large number of sensors will be essential for operation and this will make it mandatory to make the sensors miniaturized. Similarly, the actuators for the future devices will have to be composed of very large number of miniaturized actuators all working in parallel. In general, this massive parallelism will be a key feature of machines of the future, and this can be done only if economic miniaturization is achieved. Thus, it is very clear that miniaturization will be essential for not only making microsized objects but for microsized devices also.

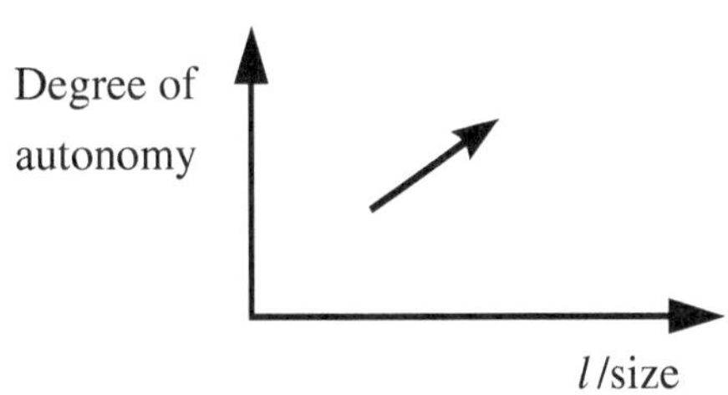

Fig. 2.3 Trend of autonomy and reduction in size of devices

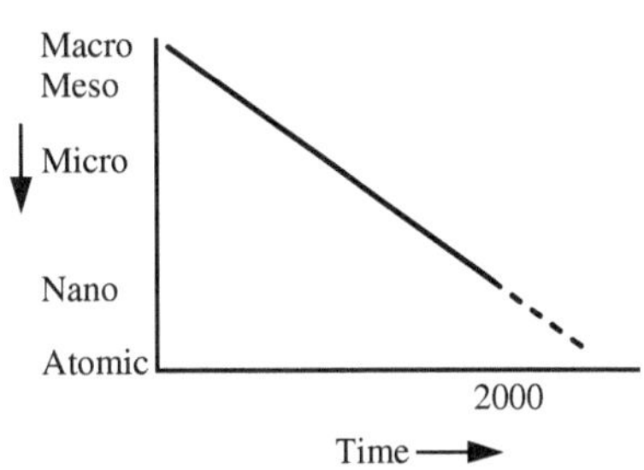

Fig. 2.4 Progress of miniaturization

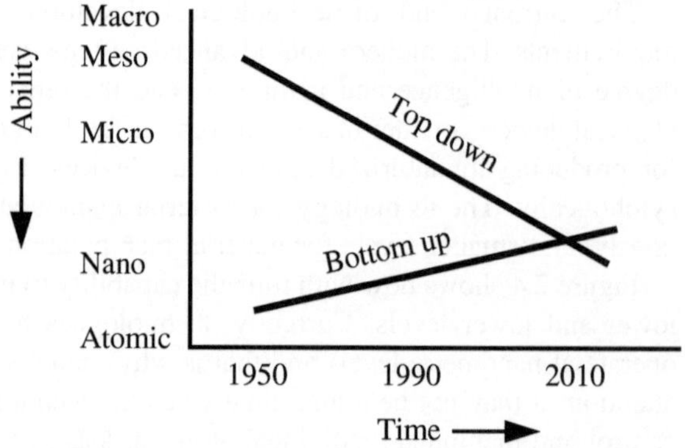

Fig. 2.5 Trends of process capabilities of top-down and bottom-up fabrication technologies

The real challenge, therefore, is the economic way to fabricate miniaturized devices. It is expected that a major shift may be experienced by the basic approach. In many situations, the technologists adopt 'bottom-up' approach along with 'self-assembly' of material for microfabrication in place of traditional 'top-down' batch production approach. Figure 2.5 represents the trend of change in fabricating miniaturized systems through the new capability for manipulating material at micro, nano and molecular levels. It is also not very difficult to recognize that if only a 'self-assembly' technique is adopted, miniaturized products can be produced by 'bottom-up' approach economically.

2.1.2 Miniaturization: Advantages and Impact on Design

The necessity of miniaturization has been discussed in the previous subsection. It may be a good idea to identify the major advantages of miniaturization before proceeding further at this stage.

One major advantage of miniaturization is the drastic reduction in cost. Though sometimes one may think reduction of size is the real important matter, the reduction in cost influences the social impact. If the central processor chip of a personal computer were ten times bigger in area, not much would have changed, but if the cost of the unit were ten times more, computer technology (or mobile communication) would have hardly been able to invade common households. Of course with reduced size, a very large number of units can be packed in a small area (or volume) which is essential for imparting intelligence to some degree.

Miniaturization also drastically reduces the consumption of energy and material. This helps in better conservation of resources and reducing the adverse effect on environment. Microsized objects have very small inertia (as will be seen) because of scaling laws, and their response is fast and can even reach the level of electronic systems.

Miniaturization is excepted to have an extremely important impact in health care technology. In many situations, revolutionary solution can be achieved in invasive health care, drug delivery, diagnostics and prosthesis. It is also very convenient for the miniaturized devices to be disposed of.

Finally, on a more subtle level, miniaturization can lead, in some cases, to taking advantage of the new laws of physics and chemistry which starts playing a predominant role in the small scales due to scaling effect.

Miniaturization will substantially influence various aspects of futuristic systems, and these needs to be considered while designing and fabricating futuristic systems involving miniaturized devices. These are represented below in a classified manner:

- *Configuration and design*—Machines are usually multicomponent devices. At the microscopic level, it is extremely difficult and uneconomic to fabricate and assemble the miniaturized components. Besides, at the microscale, the role of friction becomes predominant when members slide against one another. To avoid both these difficulties, miniaturized mechanisms should be monolithic structures in a majority of cases. The hinges are replaced by flexure joints to provide localized compliance (Fig. 2.6).
  Under some circumstances, the whole mechanism may be a compliant device and the elastic deformation of the whole unit provides the necessary movements. In cases where it is relevant, a unit may be composed of a very large number of miniaturized units all functioning is parallel.
- *Material*—Presently, Si is used for most micromechanisms as the lithographic micromachining techniques are employed to manufacture most MEMS devices. However, in more advanced systems of the future, bottom-up self-assembly-based fabrication technique may be used. In such cases, material of the unit will be typically much softer because of the requirement of self-assembly as will be seen later. Even otherwise, softer materials may be preferred in many cases for the requirement of the necessary degree of compliance.
- *Actuation*—Actuators for micromachines of the future may be very different from the conventional systems. In most cases, the actuators may be integrated with the mechanisms from the very initial stage of fabrication. In some cases, a member of the device may also act as actuator. This can be achieved by using smart materials for the elements of a micromechanism. Many exotic new ideas for microactuators are emerging, and out of all these, the concept of using molecular motors looks very promising. A number of ideas are being taken from the living world for actuating microdevices of the future, and some progress has

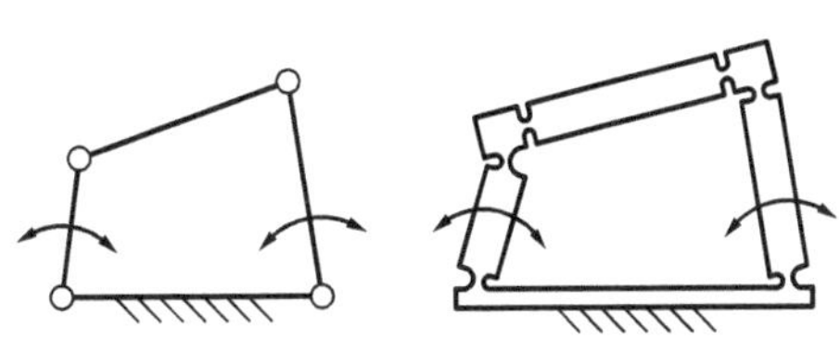

Fig. 2.6 Replacement of hinges by flexure joints

been made in developing artificial muscle type of material to drive microsystems.

- *Sensing and control*—A major difference with the conventional systems in this regard is the fact that in many situations, sensors will be integrated with the functional elements of a micromechanism. Some smart materials themselves can act as transducers. So an element in case of advanced micromechanism can have multiple tasks to achieve.
- *Energy source*—Energy source also can be of very different nature in the futuristic advanced microdevices. Unlike the conventional macroscopic systems, the energy source may not be isolated and limited in number. The energy source can be distributed over the system. One important consideration will be to reduce the distance between the energy source and the actuators. Though, till now, electrical energy is the chief source of energy, thinking has started if biochemical source of energy can be integrated with microdevices. Though it sounds like science fiction, but in future, it may be possible to operate microdevices using photosynthesis!
- *Fabrication*—Design of a system is always closely linked with the fabrication process involved. In case of micromechanisms, the fabrication technique is lithography-based micromachining. Thus, currently most such devices are primarily of 2 dimensions with thickness varying between 50 and 200 μm. Surface micromachining can make parts with mobility. More sophisticated techniques like self-assembly are still to come to reality except in a very few special situations. Some other types of fabrication processes based on generative principles may be also employed in special situations. For developing features of nanometre size, focussed ion beam (FIB) machining can be also used with which 7-nm-level feature size can be achieved. However, any process that cannot be adapted to batch manufacturing cannot be very popular because of the cost considerations.

2.2 Microsystems

Traditionally, micromechanisms represent those devices which enable some links (or elements in general) to undergo prescribed movements when input signal is applied. This movement accomplishes the desired task that can be of immense variety. The input movement is generally provided by a device called actuator as mentioned in a previous section. However, in some cases with the use of smart materials, this input motion can be generated by a member of the mechanism itself. Under many situations, the movements and associated functions of a micromechanism are controlled by signal generated by some sensing elements. In some cases, the objective of a sensor is to activate the mechanism for performing the assigned task, and in some situations, the operation of the mechanism requires feedback control, the feedback being generated by the sensors. In any case, like actuators,

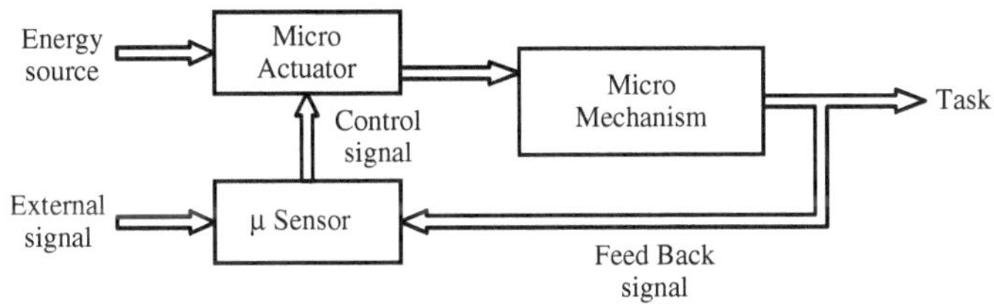

Fig. 2.7 Basic scheme of a microsystem

sensors also form an essential part of a microsystem as indicated in Fig. 2.7 symbolically. This diagram represents the structure of a typical microsystem.

Rarely, it is found desirable (or, even feasible) to totally delink the various elements of a microsystem. It is difficult to assemble all the different elements at microscale, and the system cannot be mass or batch manufactured. Besides, such separate existence may not ensure reliable functioning of the whole system under certain circumstances. Thus, the trend is to integrate all the elements forming the microsystem. These systems are so designed and the materials are so chosen that the whole system becomes monolithic and integrated. Furthermore, the design is done in a manner so as to facilitate the microfabrication of the whole unit simultaneously. In recent times, smart materials are finding increasingly more frequent usage and these materials do not only form the basic body of the micromechanism, but the same bodies can act as the actuating elements. In a truly advanced situation, the members can also play the role of the sensor elements. Such total integration provides better robustness, higher reliability and lower cost.

As the time progresses, the engineers and scientists are becoming increasingly aware of the excellent lessons from the nature, particularly the living world. This has led to a new concept in engineering—biomimetics that is based upon, imitating the biological world. In a substantial number of cases, engineers designing micromechanisms find excellent solutions to their problems from the study of insects (the branch is called entomology). One can design efficient microrobots or micromachines learning from the insect world about their methods of locomotion and movements, manipulation of objects and sensing arrangements.

While designing micromachines, a direct downward scaling of the design of a corresponding system at macroscopic scale is generally not possible, the primary reason being the scaling effects. As the relative predominance of different physical phenomena depends on the scale, a system designed for a macroscopic scale may not function at all when miniaturized isotropically. In many cases, a designer also makes decisions using his/her macro intuition. Since this capability cannot be acquired from the experiences on day-to-day encounters, a careful study of mechanics at microscale is essential. A careful study of the scaling laws is very helpful in this regard, and the next chapter of this volume is devoted to that purpose.

Chapter 3
Scaling Laws: Science of Miniaturization

Abstract The importance of different physical laws at different scales is discussed, and its relevance to the design of systems at microscopic scale is elaborated. Scaling of different types of forces is presented. The scaling laws for electromagnetic phenomenon and their relevance are explained as this plays an important role in the operation of electrical microactuators. The scaling laws for different types of flexure joints are presented as these play the role of basic building blocks for microsystems.

Keywords Scaling laws · Scaling in mechanics · Scaling of forces · Scaling laws for flexures

3.1 Scaling and Scaling Laws

Scaling laws are relationships of the parameters of a system with the size scale of the system. For example, the mass of an object made of a given material depends on its volume and, therefore, scale as length^3. On the other hand, the surface area of the same object scales as length^2. Thus, when the size of an object reduces, the mass drops at a much faster rate than its surface area. So at very small scales, those phenomena which depend on the surface area start predominating the scene. The phenomenon of scaling is known since a long time, and Galileo was the first to make methodical study of the scaling effects. However, the emergence of 'scaling laws' as a branch of serious study is relatively a recent happening. This is mainly due to the contemporary trend of miniaturization in modern technology. The knowledge of scaling laws enables a designer to predict the behaviour of either a very large system or the system at micron scale by investigating the characteristics of a laboratory model. A good understanding of scaling phenomenon and the scaling laws is, therefore, essential for successful design of micromachines and micromechanisms.

Scaling can be primarily of two types. When all aspects of a system scale in a similar way, the geometric integrity is maintained and the look of the device

A. Ghosh and B. Corves, *Introduction to Micromechanisms and Microactuators*, Mechanisms and Machine Science 28, DOI 10.1007/978-81-322-2144-9_3

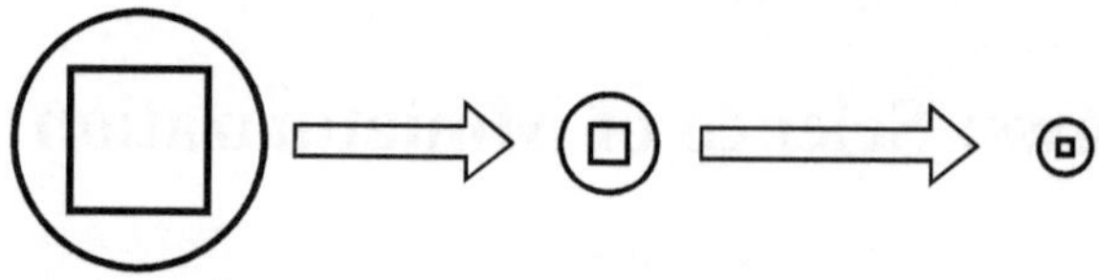

Fig. 3.1 Relative importance of phenomena at different scales

remains unchanged. This is called 'isometric' (or, 'isomorphic') scaling. But, on the other hand, when different elements of a system with different functionalities do not scale in a similar way, the shape and the look of the device may change with the scale. Such type of scaling is called 'allometric' scaling. The structural and functional consequences of change in size are dealt with by the scaling laws; thus, only through a proper understanding of the scaling laws, a designer can predict the effects of miniaturization of a device. As the size of a device is scaled down, different physical phenomena predominate at different scales as indicated symbolically in Fig. 3.1. The change in relative importance must be kept in consideration for successful operation of a miniaturized system.

As one notices, microsized specs of dust particles float in moving air, whereas bigger pieces made of the same material (i.e. stone pebbles) readily drop under gravity. This is because of the disproportionality rapid decrease in weight, whereas the air current drag, which depends on the surface area, reduces at a slower rate. As a result, at minute sizes, the drag force dominates and the effect of weight becomes insignificant. It is also a common experience that water readily spills over when a glass of water is turned upside down. But water in a capillary tube does not come out when the tube is held vertical! This is again a scaling effect that makes capillary force more prominent and the deciding factor when size is very small.

When a system is very complex, scaling laws can be very useful for understanding of the interplay among various physical phenomena and geometric features. In some extremely complex systems, the scaling laws are very simple; in many cases, these simple scaling laws can provide a much deeper understanding of some very fundamental aspects of the system. In other words, scaling laws are not only indispensible for designing miniaturized systems but are also very useful in understanding the basic principles involved in some very complex phenomena. Some examples will be presented towards the end of this chapter.

3.2 Geometric Scaling

Geometric scaling laws are well known and simple. If the shape remains unchanged, then it is known that

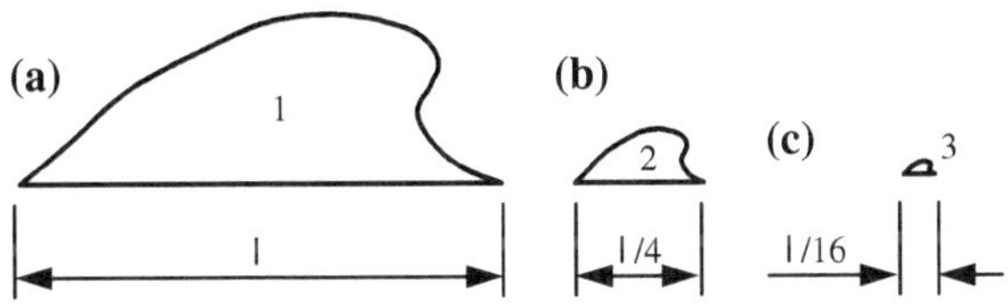

Fig. 3.2 Isometric scaling of geometric shapes

$$P \propto l$$
$$A \propto l^2$$
$$V \propto l^3$$

where P, A and V are perimeter, area and volume and l represents a characteristic length. Figure 3.2 shows three figures with identical shape but of different sizes.

So the perimeter and area of three figures are in the proportions shown below:

$$P_1{:}P_2{:}P_3 = 1{:}\frac{1}{4}{:}\frac{1}{16}$$

and

$$A_1{:}A_2{:}A_3 = 1{:}\frac{1}{16}{:}\frac{1}{256}$$

Variations of P and A with size are shown in Fig. 3.3 by graphs.

From Fig. 3.3, it is clear that those characteristics which scale as higher powers of the size become insignificant at small scales. Interestingly, the well-known Pythagoras' theorem can be stated in a more generalized way. Figure 3.4 shows a right-angled triangle ABC with three geometrically similar but irregular figures on the three sides.

As area scales as l^2, it is known that $(AB)^2 = (BC)^2 + (CA)^2$. The three figures' areas will also satisfy the relation below:

Fig. 3.3 Dependence of perimeter and area with scale

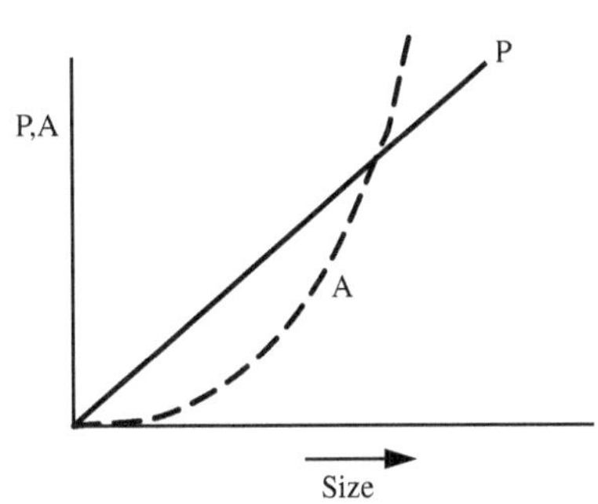

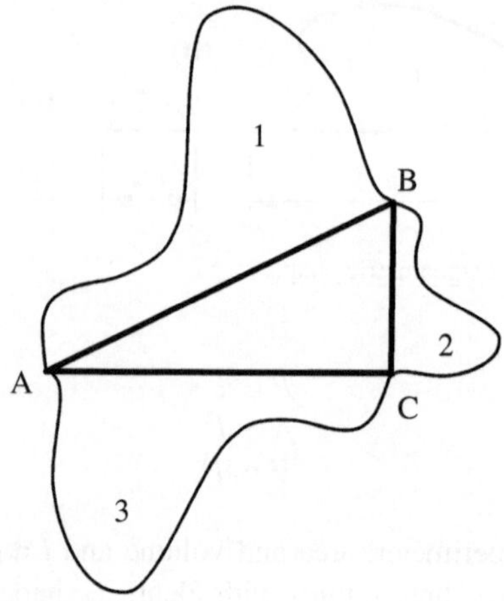

Fig. 3.4 Generalized Pythagoras' scale

$$A_2 + A_3 = A_1$$

The scaling of perimeter versus area depicts itself through many well-known observations. When thin boards are placed on water, the very small-sized pieces are found to float whereas the bigger pieces sink.

Figure 3.5a shows a square (with sides l) piece of a board that weighs w per unit area. If s be the force (upward) due to surface tension along the perimeter, then total upward force along the perimeter is equal to $S = 4ls$. At the same time, the weight of the piece (actually downwards) is given by $W = wl^2$ (where w is the weight of the board per unit area). In Fig. 3.5b, S and W have been plotted for increasing l. It is seen that as W increases as l^2 soon it will exceed S as it scales as l. Once $W \geq S$, the piece of the board sinks. For $W \leq S$, obviously the piece floats in accordance with the daily experience.

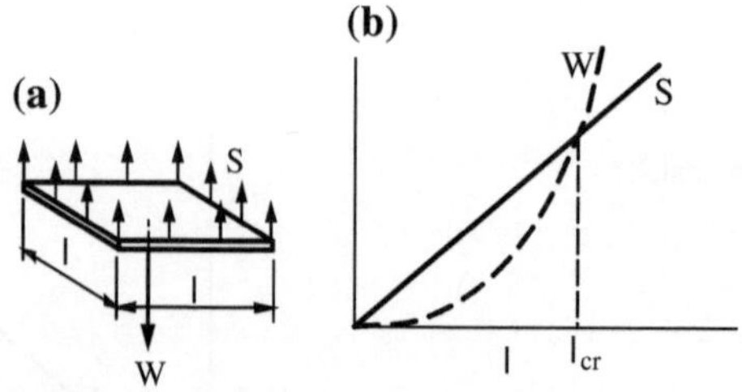

Fig. 3.5 Scale dependence of floating ability. **a** Surface tension force and weight on a square piece. **b** Growth of surface tension force and weight on size

3.3 Scaling in Mechanics

While designing mechanical systems, scaling in mechanics plays an extremely important role and must be understood well by the designers. To start with an example from solid mechanics can be considered. First the case of bending of a simple cantilever beam is considered and shown in Fig. 3.6a under the action of a concentrated force F.

It is known that the tip deflection of the beam δ is given by the following relation.

$$\delta = \frac{FL^3}{3EI} \tag{3.3.1}$$

With a rectangular cross section of the beam with b as the breadth and h as the height, $I = bh^3/12$. Thus, the equivalent stiffness of the beam

$$k = \frac{F}{\delta} = \frac{Ebh^3}{4L^3}. \tag{3.3.2}$$

As $b \propto l$, $h \propto l$ and as $L \propto l$, $k \propto l$. So the stiffness of a cantilever beam scales as l^1.

Another interesting phenomenon is the strength of a beam in relation to its own weight (Fig. 3.6b). If the density of the beam material be ρ and σ_u be the ultimate stress the beam material can withstand, then the condition of failure is given by the following equation

$$\frac{3\rho g L^2}{h} = \sigma_{max} \geq \sigma_u \tag{3.3.3}$$

where σ_{max} is the maximum stress developed at the fixed end. Thus, it is seen that the maximum stress developed due to self-weight scales as l. Figure 3.7 shows the condition for breaking under self-weight. This corroborates the experience that bigger objects of identical shape behave weaker than the smaller models. It can be

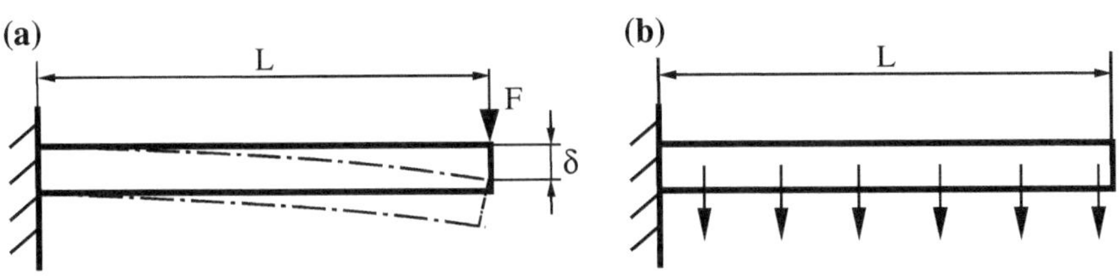

Fig. 3.6 Bending of cantilever beams. **a** Bending due to a concentrated force. **b** Bending due to self-weight

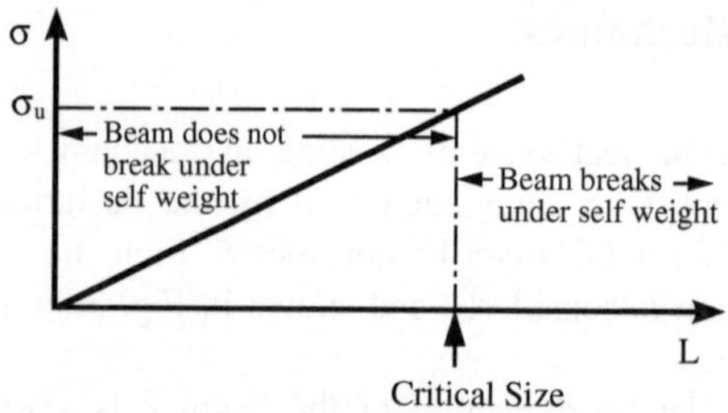

Fig. 3.7 Condition of braking of cantilever beams under self-weight

easily shown that in case of simply supported beams, also, the maximum stress developed due to self-weight $\sigma_{\max}$ scales as l.

It has been already mentioned before that with the change in size, different phenomena can take the predominant role. For example, the case of a column supporting a spherical object (as shown in Fig. 3.8) may be considered.

It is found that with the change in scale, the failure mechanism may change from compressive failure to buckling. The load F_a on the column is the weight of the spherical object and, so, F_a scales as l^3. On the other hand, the buckling load of a column with upper end free is given by $F_b = \pi^2 EI/4h^2$. This clearly shows that the buckling load F_b scales as l^2. Figure 3.9 shows the failure mechanism of the column to change from compression to buckling as the length scale (may be represented by h) reaches a critical value (say given by h_{cr}).

For dynamical systems, the scaling effects can be established using the laws of motion which involves mass, velocity and acceleration. Displacement scales as l, velocity as lt^{-1} and acceleration as lt^{-2}, whereas mass (for a given material) scales as volume, i.e. l^3. Since according to the law of motion, force is given by mass times acceleration force scales as $l^3 x l t^{-2}$, or force $\propto l^4 t^{-2}$. Work and energy are given by force times displacement, and so, they scale as $l^5 t^{-2}$. Moment of inertia

Fig. 3.8 A column support for a spherical object

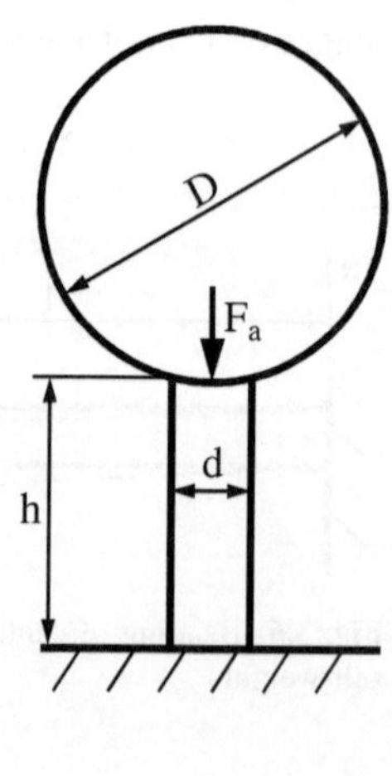

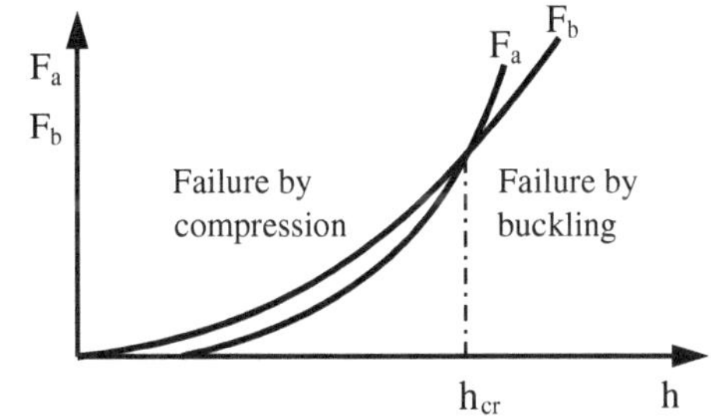

Fig. 3.9 Failure modes of a supporting column

is given by mass times length square, and it scales as $l^3 \times l^2 = l^5$. Since the moment is given by force multiplied by arm length, it has to scale as l^5t^{-2}. It is clearly seen that mass, moment and moment of inertia scale as l^3 and l^5. These parameters play very little roles at microscales.

Dynamics of elastic bodies also show dependence on scaling. If the example of a freely vibrating cantilever prismatic beam is considered, the natural frequency of transverse vibration is given by the following relation.

$$\omega_n = \sqrt{EI/A\rho}\Big/L^2 \tag{3.3.4}$$

where I is the second moment of area of the beam cross section, A is the cross sectional area, ρ is the density of the material with a modulus of elasticity E and L is the length. Thus, if the material is kept unchanged, then ω_n scales as l^{-1}. Therefore, with reduction in size, ω_n increases and microsystems possess natural frequencies in the MHz (even GHz in extreme cases) range. As a result, the response time for micromechanical devices can be comparable to that in electronic systems.

The effect of scaling laws is more pronounced in case of fluid flow related systems. In microchannels, the Reynolds's number (Re) for a fluid flow is very low and can be even around 1 as can be shown from the fact that $Re \propto l^2$. Flow with such law Reynolds' number is extremely laminar and, as a result, it is extremely difficult to mix fluids in microchannels. It is further known that for a fluid flow through a pipe of radius r and length L, the volume rate of flow (Q) and the pressure difference at the two ends (Δp) satisfy the following relation

$$Q = \left(\frac{\pi r^4}{8\eta L}\right)\Delta p \tag{3.3.5}$$

when η is the viscosity of the fluid. From (3.3.5), one gets the following equation

$$\left(\frac{\Delta p}{L}\right)\Big/Q = \frac{8\eta}{\pi r^4}$$

Hence, for a given volume flow rate through a pipe, the pressure drop rate ($\Delta p/L$) scales as l^{-4}. So it is extremely difficult to push fluids through pipes of small diameters. In fact, different mechanisms are used in micropumps for this reason.

3.4 Scaling in Electromagnetism and Electrostatics

Since actuating devices depend heavily upon electrical systems, it is desirable to have some idea about the way the scaling effects influence electrical phenomena. To start with the effect of scaling on three common elements in electrical circuits is considered. The resistance of a given conductor of length L, cross sectional area A and specific resistivity γ is given by

$$R = \frac{\gamma L}{A} \propto l^{-1}$$

So for a given material, resistance scales as l^{-1}. In case of a parallel-plate capacitor with plate area A, plate gap d and ε as the permittivity of the medium filling, the plate gap is defined as follows:

$$c = \frac{\varepsilon A}{d} \propto l.$$

Hence, capacitance scales as l. The inductance of a coil with n number of coils per unit length, coil area A and L as the length of the coil is

$$I = \frac{\mu n^2 A}{L} \propto l$$

where μ is the permeability of the medium within the coil. Thus, inductance scales as l.

When two identical parallel plates are impressed upon with a potential V, at the symmetric position indicated in Fig. 3.10, the electrostatic force components F_d, F_L and F_B can be expressed as follows:

$$\begin{aligned} F_d &= -\frac{\partial U}{\partial d} \\ F_L &= -\frac{\partial U}{\partial L} \\ F_B &= -\frac{\partial U}{\partial B} \end{aligned} \tag{3.4.1}$$

where U is the potential energy stored in the capacitor, given by $-1/2CV^2$. Since capacitance $C \propto A$ and $\propto 1/d$, $U \propto AV^2/d$. The voltage that can be applied without

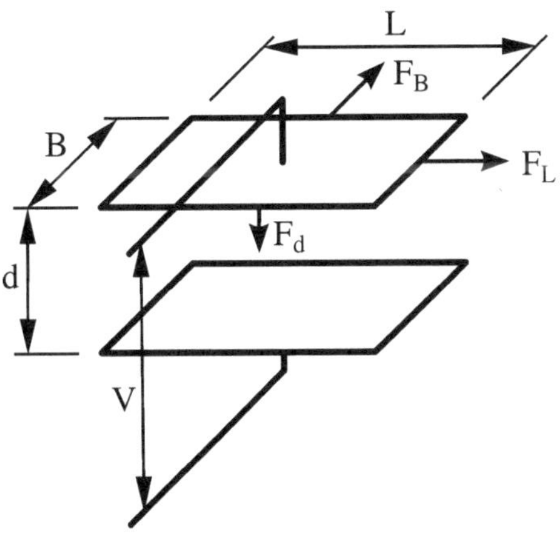

Fig. 3.10 Electrically charged parallel plates in symmetric configuration

break down (which follows Paschen's effect) is approximately proportional to the gap d, i.e. V scales as l. Therefore, the electrostatic potential energy

$$U \propto \frac{l^2 \cdot l^2}{l} \propto l^3 \tag{3.4.2}$$

Equations (3.4.1) and (3.4.2) yield the following scaling effects:

$$\begin{aligned} F_d &\propto \frac{l^3}{l} \propto l^2 \\ F_L &\propto \frac{l^3}{l} \propto l^2 \\ F_B &\propto \frac{l^3}{l} \propto l^2 \end{aligned} \tag{3.4.3}$$

Thus, in general, electrostatic force scales as l^2. To find out the scaling effect on electromagnetic force, the conductor shown in Fig. 3.11 can be considered. L is the length of the conductor carrying a current i in a magnetic field B as shown.

The force F acting on the conductor is given by the following equation:

$$F = iLB \tag{3.4.4}$$

For a prescribed current density, i is proportional to the cross section area of the conductor. Hence,

$$i \propto l^2. \tag{3.4.5}$$

Again,

$$B = \mu_0 ni/L \tag{3.4.6}$$

where, n is the number of turn in the coil and μ_0 is the permeability. So, B scales as l. Thus, finally the above three relations result in

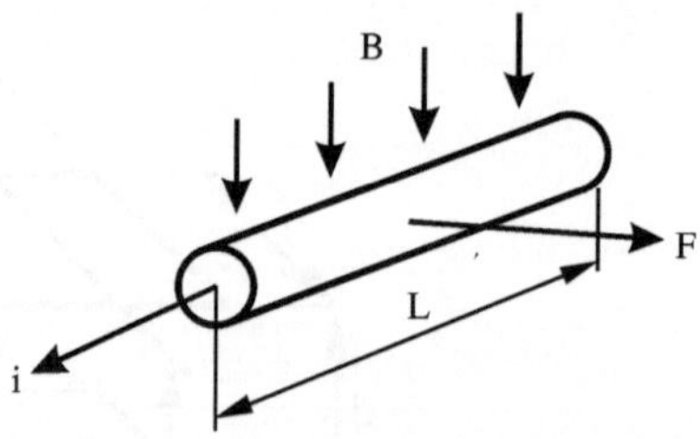

Fig. 3.11 Electromagnetic force on a conductor in a magnetic field

$$F \propto l^4 \tag{3.4.7}$$

or, electromagnetic force scales as l^4. Because of these scaling laws, it is clear that at small scales, electromagnetic actuators are less effective than electrostatic actuators as it scales as l^2. This matter has great importance for designing actuators for microsystems.

3.5 Scaling of Common Forces

Movement of devices and systems are caused by forces according to the laws of motion. Therefore, an idea about the scaling of common type of forces should be known. Table 3.1 presents the scaling of the common forces.

3.6 Scaling in Micromechanisms

In an earlier section, it has been emphasized that at microscale, it is extremely difficult to assemble microparts for fabricating micromechanisms. Therefore, most often the micromechanisms are created by monolithic structures with lumped compliances where the joints are expected to be present. Commonly, these lumped

Table 3.1 Scaling different types of forces

Force type	Scaling
Surface tension	l^1
Electrostatic force	l^2
Fluid force	l^2
Weight/inertia	l^3
Electromagnetic force (for constant current density)	l^4
Van de Waal's force	l^1

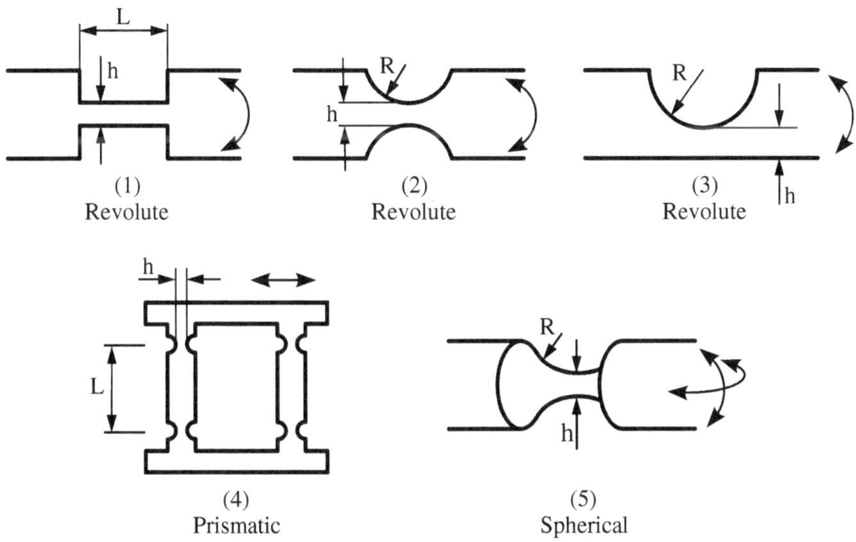

Fig. 3.12 Different types of flexure hinges

Table 3.2 Scaling effect for different types of flexure hinge

S. no.	Simulates	Maximum deflection for a given material (limiting stress σ) ($\delta_{\max}$)	Scaling effects
1	Revolute	$2\sigma L^2/3Eh$	$\propto l^1$
2	Revolute	$(3\pi\sigma/4E)\sqrt{R/h}$	$\propto l^0$
3	Revolute	$(3\sqrt{2}\pi\sigma/4E)\sqrt{R/h}$	$\propto l^0$
4	Prismatic	$(3\pi\sigma/16EL^2)\sqrt{R/h}$	$\propto l^2$
5	Spherical	$(5\pi\sigma/8E)\sqrt{R/h}$	$\propto l^0$

compliances are produced by creating locally compliant sections. Figure 3.12 shows the common types of these flexural hinges used in developing micromechanisms. The first three types of flexural hinges are used to simulate revolute joints. The fourth one is for prismatic joints, and the last one is to present spherical joints.

Table 3.2 represents the deflection characteristics $\delta_{\max}$ of these commonly used flexural hinges. It is seen from the expression for the $\delta_{\max}$ that except for the first and fourth cases, $\delta_{\max}$ is independent of the size. But in the first and the fourth cases, these flexural hinges tend to behave much stiffer at small scales.

Chapter 4
Micromechanisms

Abstract The chapter starts with a discussion on how the emergence of micromechanisms and the realization of its importance in the impending paradigm shift in technology have revived the interest in the subject 'Mechanism'. Assembled micromechanisms and compliant micromechanisms are presented along with a general discussion on various characteristics of miniaturized devices.

Keywords Micromechanisms · General characteristics · Assembled micromechanisms · Compliant micromechanisms

4.1 Micromechanisms and Revival of Mechanism Theory

Theory of mechanisms and machines has always been a hard core traditional mechanical engineering subject. This branch of study was considered to be absolutely essential to all mechanical engineering students. The primary theme of the subject used to be the technique of generating desired motions of objects other than just pure rotation. Generating complicated motion with stipulated accuracy requirement required special knowledge about synthesizing mechanical systems generally called mechanisms (or machines). However, with the advent of control engineering, electronics and computer technology complicated movements could be generated without using complex mechanisms as illustrated in Fig. 4.1.

In Fig. 4.1a, a complex curve is being generated by a coupler point of a 4R mechanism; on the other hand, such a complex path can be generated by controlling the *X–Y* motions using suitable programming as indicated in Fig. 4.1b. The second approach can not only generate any complex path but can do that with high accuracy which is not generally possible by a mechanism. As a result, the interest in studying a subject devoted to synthesis and design of complex mechanisms gradually declined, and in many places, these courses were removed from the curricula.

Fortunately, the interest in the subject of the synthesis and design of mechanism was revived with increasing population of micromechanisms. Since generating

A. Ghosh and B. Corves, *Introduction to Micromechanisms and Microactuators*, Mechanisms and Machine Science 28, DOI 10.1007/978-81-322-2144-9_4

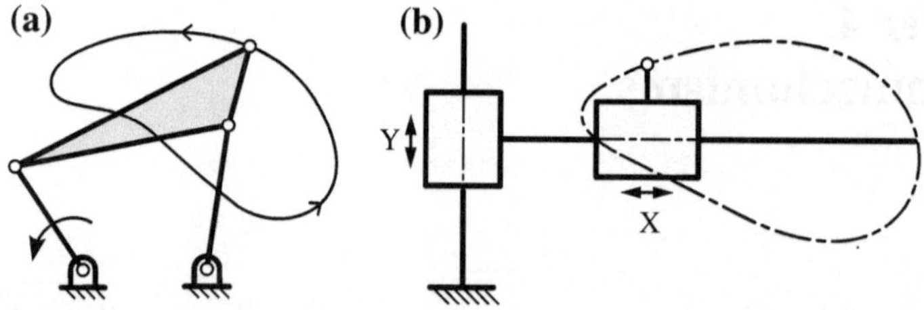

Fig. 4.1 Methods for generating a curve. **a** Using coupler curve of a constrained mechanism. **b** Using coordinated movement

complex motion at the microscopic level using controlled individual drives is not practicable, engineers had to revert back to mechanisms (though at microlevel). The renewed interest in mechanism theory is, therefore, a direct consequence of the emergence of micromechanisms and microactuators as important elements of modern technology.

There are a number of reasons for increasing popularity of microsystems as identified below:

1. Cost and size both decrease with miniaturization and, therefore, advanced intelligent systems can be developed using very large number of microdevices working in a massively parallel mode.
2. Non-invasive health care and intelligent drug delivery systems become possible.
3. Better conservation of resources results from low material and low energy consumption.
4. Response of the devices at microscale is very fast.
5. Microsystems are easily disposable.
6. New laws of mechanics, physics and chemistry at very small scales can satisfy the requirements of a particular task.

The application areas of micromechanisms and microdevices can be identified as follows:

7. Medical and life science.
8. Space-related technology and applications.
9. Advanced electronic and photonic devices.
10. Development of highly intelligent autonomous systems.

It is also appropriate to have an overall idea about different aspects of micromechanisms as shown in Table 4.1.

Table 4.1 shows the important fact that at extremely small scales, externally motivated material manipulation for fabricating micromachines is no longer possible and the technique of molecular (or nanoparticle) self-assembly has to be adopted. So far as the basic components of micromachines are concerned, Table 4.2 shows the trend in a consolidated format.

Micromechanisms consist of mechanical elements of submillimetre dimensions with corresponding tolerance in the submicron range. In fact, the micromechanisms

Table 4.1 Some characteristics of miniaturized devices of different scales

Systems	Size range	Observation method	Components	Manipulation & Fabrication
Milli machines	10 mm 1 mm	Naked eye	Miniature components	Precision fabrication
Micro machines	100 µm 10 µm 1 µm	Optical microscopy	Micro components	UV lithography & μ machining
	100 nm		Nano components	X-Ray lithography
Nano machines	10 nm 1 nm	Electron microscopy	Molecular components	Protein engineering and molecular self assembly
	0.1 nm			

which occupy dimensions in the millimetre (or submillimetre) range may be strictly called 'millimachines' and when the dimensions are in the micron range, the term 'micromechanism' appears to be appropriate. However, both these groups are often combined under the common heading 'micromechanisms'. Micromechanisms are used for both manipulation at the submillimetre or microlevels and also for micropositioning which is very important in many advanced technologies like scanning electron microscopy, X-ray lithography, mask alignment, advanced optical device, etc. Micromanipulation technology has made revolutionary progress in life science research involving cells and such microobjects.

4.2 Assembled Micromechanisms

Some micromechanisms are composed of a number of moving bodies (like a typical mechanism in the macroscopic scale). Techniques of precision fabrication and then suitable assembly result in the final mechanism. However, as expected, such systems cannot be reduced to microscale and, even otherwise, the cost of such devices tends to be high because of the involved assembly. An example of this type of micromechanism is shown in Fig. 4.2.

Table 4.2 Aspect of normal and micromachines

Aspects	Normal Machines		Micromachines
Topological feature	Multicomponent	→	Monolithic
Material	Hard and dumb	→	Soft and smart
Sensors	Microscopic/mesoscopic	→	Microscopic & nanosized
	Limited numbers	→	Very large numbers
	Separate	→	Integrated
Actuators	Electric, hydraulic, pneumatic motors and actuators	→	Smart materials, ionic polymer based artificial muscles and molecular motors.
	Centralized	→	Distributed
Energy source	External and predominantly electrical	→	Internal and electrical, biochemical in future
	Centralized	→	Distributed
Manufacturing	Conventional top down process	→	Lithography
		→	Self assembly based bottom up processes

The vibratory motion of the tuning fork (generated by an oscillating current in the electromagnet) causes the pawl to move back and forth and, in turn, rotate the ratchet wheel. Electrostatic motors described earlier also consist of individual moving member. Piezoelectric micromotors, microengines and microgasturbines all have been developed successfully. Microrobots are also being developed for many different types of applications.

Producing devices at still smaller scales with multiple components has been made possible by surface micromachining techniques based on lithography. With surface micromachining, it is possible to produce systems with multiple parts capable of making relative motion. With surface micromachining, the assembly step is eliminated making extreme miniaturization feasible in a practical sense. Microsurgery is another important area of application of microdevices. Although these exotic

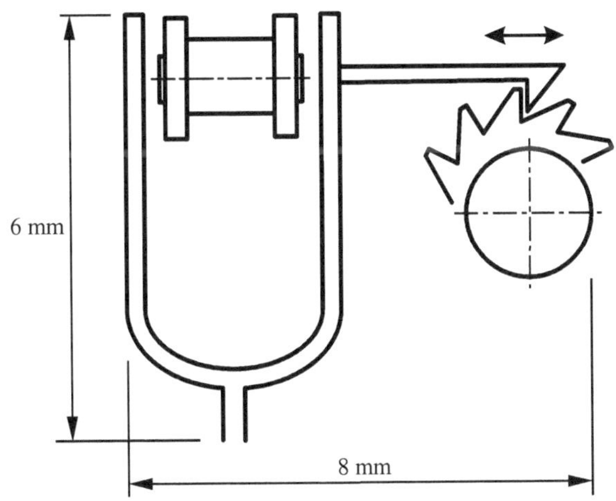

Fig. 4.2 A multicomponent micromechanism

products are very interesting and are quite useful in many situations, they suffer from a serious problem apart from being expensive. Because of the very high surface-to-volume ratio and low rigidity, large interfacial attractive forces quite often create the problem of stiction. Stiction is a phenomenon that results in strong (and often permanent) attachment of contacting elements. The scaling effect renders capillary, van der Waal's and electrostatic attractive forces quite predominant.

Such problems of microassembly, friction and wear have encouraged technologists to conceive micromechanisms as monolithic structures capable of performing the tasks of a multicomponent device.

4.3 Monolithic Systems: Compliant Micromechanisms

It was mentioned earlier that limited relative movement among various sections of a monolithic structure can be achieved by introducing localized compliance at suitable sections. Figure 4.3a shows a link with a localized weak portion C in the form of a deep notch. When subjected to bending moment, the link will take the shape as shown in Fig. 4.3b.

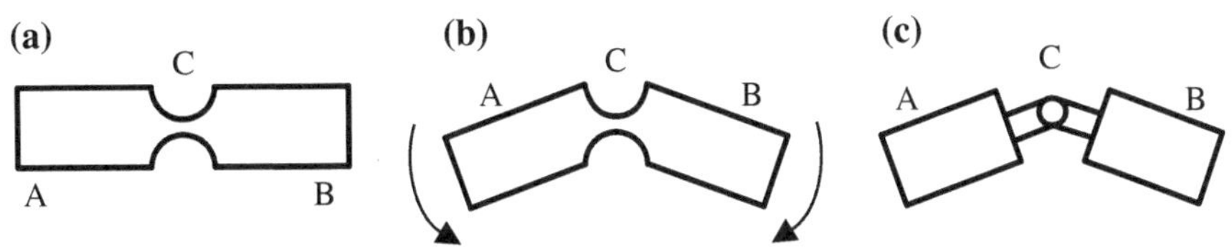

Fig. 4.3 Simulation of a hinge by a flexure point. **a** Undeformed flexure. **b** Flexure joint in bent condition. **c** Equivalent configuration of a hinge joint

Most of the deformation is localized in the weak part C, whereas the rest of the link behaves like rigid bodies undergoing very little deformation. The result is as if the two parts A and B are connected by a revolute joint at C, as indicated in Fig. 4.3c. Most of the elastic energy will also be stored in the flexible section of the link. This can be easily seen from a somewhat analogous situation where three linear springs with stiffnesses k_{A}, k_{B} and k_{C} are connected in series.

If a force F is applied at the free end, the elongations of the three sections will be as follows:

$$\delta_{\mathrm{A}} = \frac{F}{k_{\mathrm{A}}}$$

$$\delta_{\mathrm{B}} = \frac{F}{k_{\mathrm{B}}}$$

$$\delta_{\mathrm{C}} = \frac{F}{k_{\mathrm{C}}}$$

Now as in the case of the link with the localized compliance, if k_{A} and k_{B} be much higher than k_{C}, then obviously δ_{A} and δ_{B} will be much smaller than δ_{C}. The movement of the point of application of the force F, B, will be almost equal to δ_{C}, the other two extensions being negligible. The energy stored in the two tough springs will be equal to $(1/2)F\delta_{\mathrm{A}}$ and $(1/2)F\delta_{\mathrm{B}}$ which will be negligible compared to the energy stored in the central spring as δ_{A} and $\delta_{\mathrm{B}} \lll \delta_{\mathrm{C}}$. This phenomenon has been used by designers to design micromechanisms as monolithic structures with localized compliances wherever revolute joints are supposed to exist in a corresponding conventional mechanism. Of course, the equivalence between a revolute joint and a flexure joint is not exact but those details can be discussed later. For simple non-precise manipulation, such compliant mechanisms can provide satisfactory performance. Many types of microgripping devices have been designed following this approach. Microsized straight line generating devices have also been successfully developed using such flexure hinges.

There is another class of micromechanisms which are monolithic structures but the desired manipulating movements are generated by the elastic deformation of the whole body, or at least a major part of the whole structure. Such mechanisms need more sophisticated design based on finite element analysis of the complete structure. Using optimization techniques, the design process has been rendered more sophisticated as will be seen later.

Chapter 5
Design of Micromechanisms

Abstract This chapter starts with a discussion on design of micromechanisms with multiple members. The special problem of adhesion and stiction is explained from basic physics. Design of compliant micromechanisms using flexural hinges is presented, and the characteristics of different types of flexures are elaborated and compared. The approach for modelling a flexure hinge and the parasitic movements are explained. A brief outline of the strategy for designing micromechanisms with distributed compliance ends the chapter.

Keywords Flexure hinges · Distributed compliance · Structural architecture of micromechanisms · Stiffness characteristics of flexure hinges · Modelling of flexure hinges

5.1 Design of Micromechanisms with Multiple Members

Development of micromechanisms with several microparts requiring assembly faces serious challenges, particularly from the point of assembly of microparts. However, the emergence of the surface micromachining technology using sacrificial layer technique has made the fabrication of multiple member microsystems economically and technologically feasible. It is possible to design a mechanism following the kinematic synthesis procedure followed by mechanical design process as done in the macroscale. However, there are special design issues because of the scale and fabrication process which must be kept in mind. Of particular importance are the clearances and tolerances—line width of the lithography, spacing for the sacrificial layer, and the dimensional tolerances of the mechanical elements. So far as the design of the basic configuration is concerned, the usual procedure that is followed for designing the macroscopic counterpart of similar mechanisms is adopted. But as mentioned earlier, the layout is decided keeping the surface micromachining process in mind. In many cases where electrostatic actuation is used, the problem of stiction can impede normal functioning of a mechanism unless proper care is taken from the very design stage. The phenomena of high

A. Ghosh and B. Corves, *Introduction to Micromechanisms and Microactuators*, Mechanisms and Machine Science 28, DOI 10.1007/978-81-322-2144-9_5

adhesion during pull and high static friction during sliding between two elements are referred to as 'stiction'. A considerable amount of research has been carried out to investigate the phenomena of adhesion and stiction. Concept of 'free surface energy' is quite helpful in this respect. When a crystalline solid is cleaved along a cleavage plane, highly active surfaces are generated because of the fracture of cohesive bonds which would have been used for bonding with other atoms before fracture. The energy required to create a new surface is called the 'free surface energy'. It not only determines adhesive bonds, friction, and wear but also influences the interaction of lubricants with solid surfaces. If two surfaces with γ_1 and γ_2 as the free surface energies (also surface tensions) in air are put in contact, the surface energy of the interface becomes γ_{12} (the surface energy is always per unit area). Thus, the reduction in surface energy is given by

$$\Delta\gamma = \gamma_1 + \gamma_2 - \gamma_{12} \tag{5.1.1}$$

and is always a negative quantity and is also equal to the work of adhesion (per unit area). $\Delta\gamma$ represents the energy per unit area that must be applied to separate the two surfaces.

When there is a liquid film between two surfaces, two possible situations can exist as indicated in Fig. 5.1. If the surfaces are hydrophilic (water liking), then the contact angle θ is less than 90°. On the other hand, if the surfaces are water repellent, i.e. hydrophobic, $\theta > 90°$ as shown in Fig. 5.1b.

When $\theta < 90°$, the liquid pressure p_F is less than air pressure p_A, and the equilibrium configuration is as shown in Fig. 5.2.

If an element of the fluid–air surface subtending an angle $d\phi$ at the meniscus centre O and of width ds is considered, its equilibrium requires the following relation:

$$2\gamma_{FA} \cdot ds \cdot \sin\frac{d\phi}{2} = (p_A - p_F) \cdot r\,d\phi \cdot ds$$

As $d\phi$ is very small, the above relation becomes as below:

$$\Delta p_{FA} = (p_F - p_A) = -\frac{\gamma_{FA}}{r} \tag{5.1.2}$$

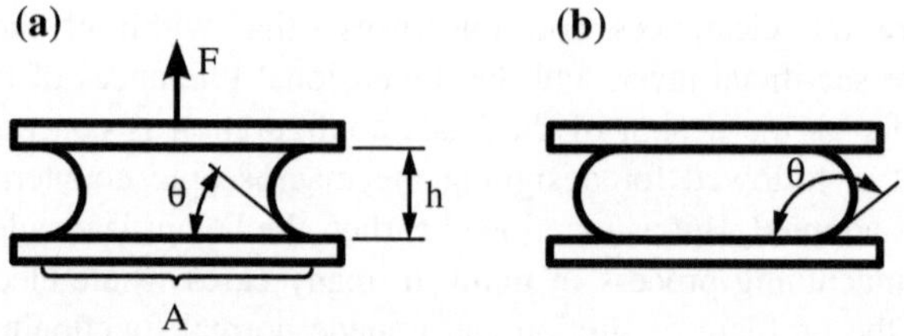

Fig. 5.1 Contact configuration of liquid films between two solid surfaces. **a** Hydrophilic surfaces. **b** Hydrophobic surfaces

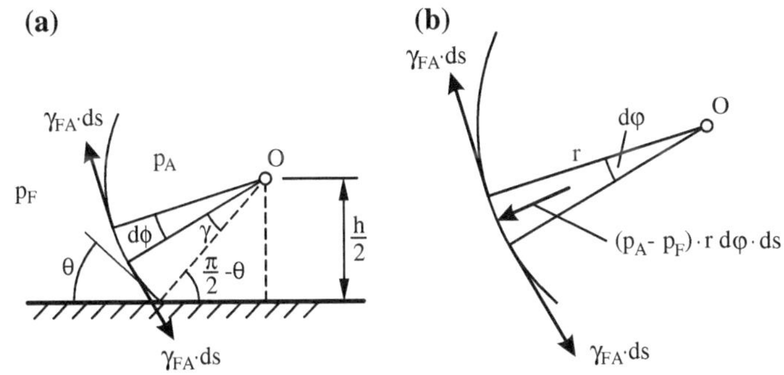

Fig. 5.2 Determination of contact angle. **a** Liquid surface configuration. **b** Equilibrium of surface element

when r is assumed to be negative for $\theta < 90°$. Again from geometry (Fig. 5.2a)

$$r \sin\left(\frac{\pi}{2} - \theta\right) = \frac{h}{2}$$

or,

$$r = -\frac{h}{2\cos\theta} \tag{5.1.3a}$$

(as r is negative by convention). Hence, the force required to separate the two plates, F, (Fig. 5.1a) is given by

$$F = -(p_{\mathrm{F}} - p_{\mathrm{A}})A = -\gamma_{\mathrm{FA}}A \tag{5.1.3b}$$

where A is the area of fluid contact. Combining (5.1.2) with (5.1.3a, b), one gets

$$F = \frac{2A\,\gamma_{\mathrm{FA}}\cos\theta}{h} \tag{5.1.4}$$

When at equilibrium, θ is determined by the surface tensions γ_{FA} and γ_{SF} as shown in Fig. 5.3.

$$\gamma_{\mathrm{FA}}\cos\theta + \gamma_{\mathrm{FS}} = \gamma_{\mathrm{SA}}$$

or

$$\theta = \cos^{-1}\left(\frac{\gamma_{\mathrm{SA}} - \gamma_{\mathrm{FS}}}{\gamma_{\mathrm{FA}}}\right) \tag{5.1.5}$$

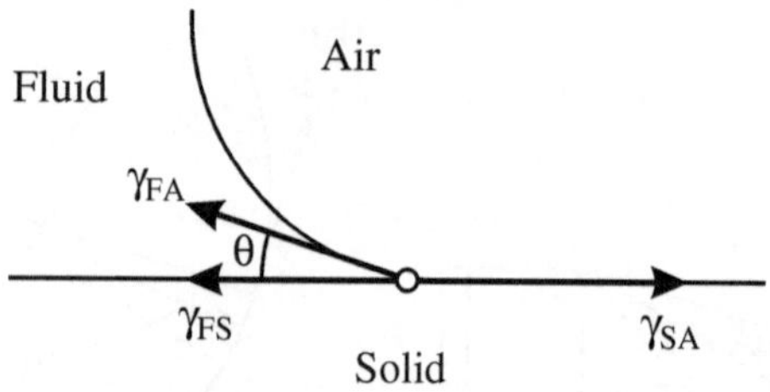

Fig. 5.3 Solid–liquid–air interface configuration

For the situation indicated in Fig. 5.1a, the resultant surface energy can be expressed as follows:

$$E = A_{tot}\gamma_{SA} - A(\gamma_{SA} - \gamma_{SF})$$

where A_{tot} is the total surface area of the solid face and A is the area in contact with the fluid. In case of micromachines with moveable parts, a member may get adhered to the substrate via a liquid film if the total energy reaches a minimum.

There are anti-stiction coatings that protect the system from this adherence problem. These coatings primarily manipulate the surface energy values in a desired appropriate way. Other details of micromechanism design are closely linked with surface micromachining and other lithographic processes. These will be discussed along with the fabrication processes in a subsequent chapter.

5.2 Compliant Micromechanisms

Some very interesting experimental investigations have been carried out to find out the comparative performance characteristics of multimember micromechanisms and their monolithic twins. In general, it has been noticed that performance of the multimember micromechanisms is inferior to that of monolithic compliant micromechanisms because of backlash, friction and wear. Even the fatigue life is much less than that of compliant mechanisms. Furthermore, it has already been pointed out that batch production of monolithic compliant mechanisms is much easier and it has resulted in much better productivity and easy commercialization. Thus, a tremendous growth is taking place in the area of compliant micromechanisms.

Before the design of a micromechanism is taken up, certain points need to be kept in mind. In macroscopic mechanisms, the typical values for the range of motion are 10^{-1} m and the corresponding positional accuracy is 10^{-5} m. Hence, the ratio of accuracy/range is of the order 10^{-4}. If the same logic is extended for this ratio, the accuracy in micromechanisms should reach the nanometre levels. Thus, success of designing a micromechanism depends on how such high degree of accuracy can be achieved. In case of compliant micromechanisms, there are primarily two sources of inaccuracies—(i) cross-flexural and parasitic effects in the flexure hinges and

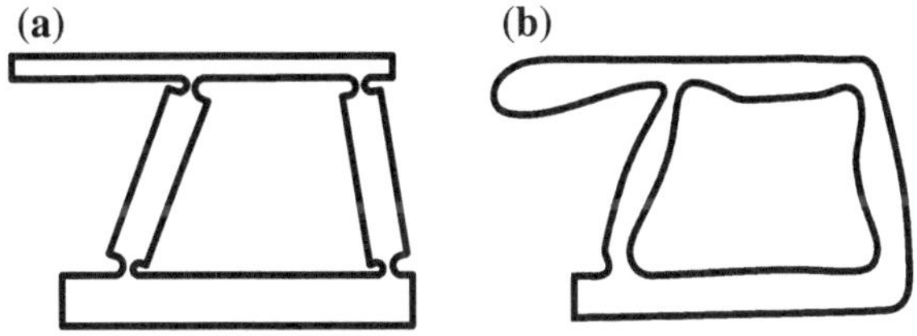

Fig. 5.4 Compliant mechanism types. **a** Lumped compliance. **b** Distributed compliance

(ii) the errors in modelling. There are primarily two types of compliant micromechanisms. One type is based on the kinematics of a corresponding conventional mechanism with multiple rigid links, and only the hinges are replaced by flexure joints. These are sometimes called as lumped compliant mechanisms. The other type of compliant mechanism is called distributed compliant mechanism. In case of micromechanisms under the later group, the motion is realized by elastic deformation of the whole structure. Figure 5.4 shows the different types of micromechanisms based on link compliance and localized flexural members.

5.2.1 Compliant Micromechanisms with Flexure Hinges

This type of systems was developed first as it is easier to get reasonable solution with a relatively simpler approach. The design process can be generally split into two stages. In the first stage, dimensional synthesis of the basic kinematic structure with rigid members (as in case of a corresponding conventional mechanism) is performed. Then, the revolute joints are replaced by the flexure hinges, which provide the very localized ability to undergo bending (or rotation). There are ways to replace prismatic joints though it is a little more involved. The second phase of the design involves a detailed elastic analysis and optimization to achieve the desired level of accuracy. A typical flexure joint to replace a hinge is shown in Fig. 5.5a.

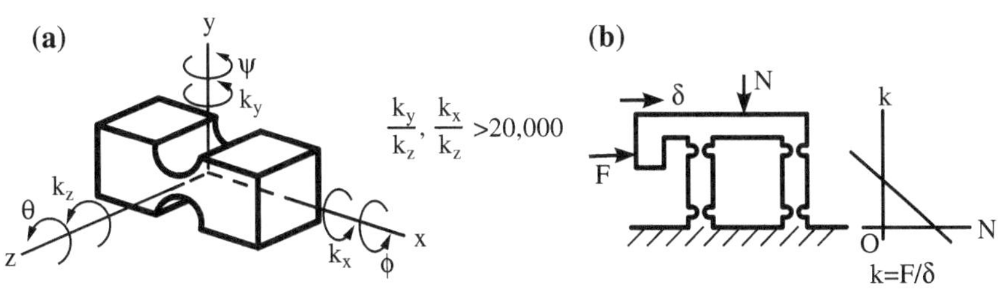

Fig. 5.5 Flexure hinge-based prismatic motion generation. **a** Typical flexure joint. **b** Prismatic joint action

The joint provides predominant bending possibility about axis z. For bending about other two axes, x and y, the bending stiffness is very high and the ratios, k_x/k_z and k_y/k_z, are more than 20,000. Hence, the flexure approximately represents a revolute joint as demonstrated in Fig. 5.6.

A combination of this flexure element can provide a capability of translational movement δ as shown in Fig. 5.5b and thus can approximately represent a prismatic joint. Figure 5.7 shows the equivalent prismatic joint. The use of flexure hinges eliminates friction, backlash, and wear.

These compliant mechanisms are sensitively dependent on feature geometry and material properties. This leads to substantial uncertainty in performance, and therefore, model validation is an important step in micromechanism design. A new branch of study has emerged for this step that is known as 'elastokinematics'. The typical design process of a compliant micromechanism is indicated in Fig. 5.8. To start with a rigid body mechanism is synthesized that generates the desired movements and by a few iterations it is finalized. In the next stage, this rigid body mechanism is gradually replaced by a compliant mechanism that reproduces the motions. This design is perfected by optimization process as indicated in Fig. 5.8. When the performance reaches an acceptable level, the final design is accepted.

Even though the use of compliant micromechanisms has become popular as the effects of friction, backlash, wear, etc., are eliminated, compliant micromechanisms are sensitive to variations in material properties and feature geometry resulting in substantial uncertainties in performance quality. Thus, model validation is an important step before the design is concluded. All realistic flexural members replacing the revolute joints differ from the ideal revolute and prismatic joints, and these elastic joints generally exhibit some cross-flexural effects. These effects deteriorate the precision of resulting motion. These parasitic effects have to be taken into account in the calculations for either compensating or completely eliminating these effects. Selecting the material is another important step; materials with high

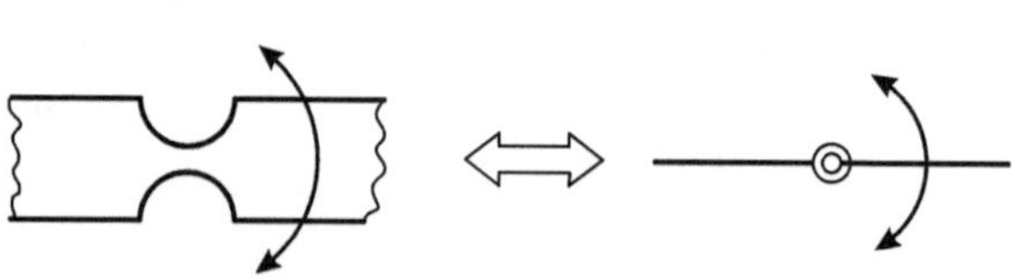

Fig. 5.6 Mobility characteristics of flexure joints and hinges

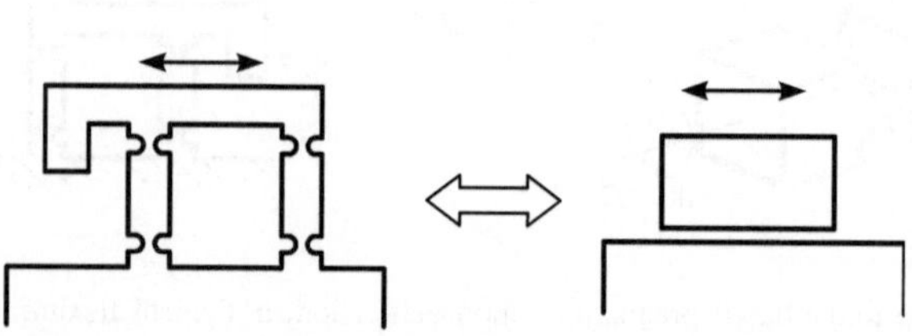

Fig. 5.7 Prismatic joint action 'using flexure' elements

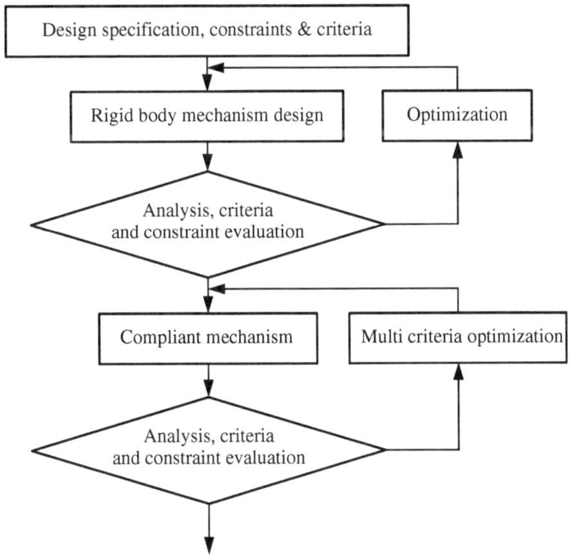

Fig. 5.8 Strategy for designing compliant mechanisms

ratio of strength over elasticity should be chosen. To illustrate the point, a few standard flexure hinges are considered and the relevant points are presented in a tabular format as presented in Table 5.1.

δ_{max} represents the maximum permissible characteristic deflection, σ is the maximum permissible stress in the material, and E is the modulus of elasticity of the

Table 5.1 Types of flexure hinges and basic characteristics

Flexure hinge type		Maximum permissible characteristic deflection	Shape
Revolute	Cross section	δ_{max}	
	Rectangular	$2l^2\sigma/3Eh$	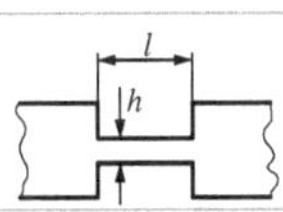
	Double notched	$3\pi R^{0.5}\sigma/4Eh^{0.5}$	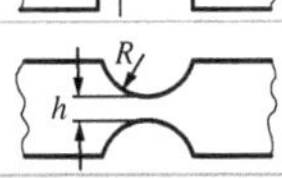
	Single notched	$3\sqrt{2}\pi R^{0.5}\sigma/4Eh^{0.5}$	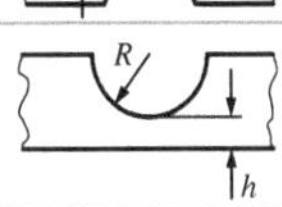
Prismatic	$3\pi R^{0.5}\sigma/16l^2Eh^{0.5}$		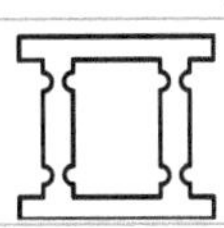
Spherical	$5\pi R^{0.5}\sigma/8Eh^{0.5}$		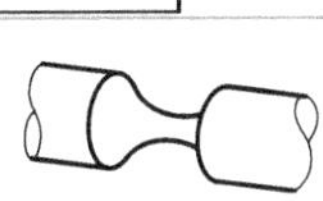

material. The objective should be to achieve maximum possible deflection in order to achieve largest mobility and workspace. It is clear from the above table that the deflection bears the following type of relation with the material properties:

$$\delta_{\max} = \lambda \left(\frac{\sigma}{E}\right). \tag{5.2.1}$$

In other words, the key material parameter is the ratio σ/E and, obviously, the attempt should be to maximize this ratio. Most often, Si is used for manufacturing compliant micromechanisms as it has excellent structural properties and it is also suitable for integration with the associated electronics. Lithography-based micromachining technology is also very well developed for Si because of the progress in the microelectronic industry. Si also possesses reasonable fatigue properties.

Another important task in the preliminary stage of the design is choosing a suitable structural geometry. The number of degrees of freedom of the mechanism has to be as per the prescribed task to be accomplished; extra degrees of freedom should be avoided as unnecessarily large number of degrees of freedom makes the control more difficult. The structure should be as simple and compact as possible with shortest possible kinematic loops. For high-precision applications, a parallel architecture is more suitable. Depending on the method of actuation, four basic types of parallel mechanisms are possible as mentioned below in tabular form (Table 5.2).

5.2.1.1 Flexure Hinges

Study and analysis of flexure hinges are very essential for designing micromechanisms with lumped compliances. Since these elements play the predominant role in the performance of lumped compliance, micromechanisms' extensive study on their design has been carried out. The primary objectives to be achieved are as follows:

- Provide large compliance
- Develop as low stress as possible
- Cause minimum deviation from a pure hinge kinematically

The shape of the notch contour is the deciding factor for satisfying the above objectives for a given material. There are some parameters which are relevant to flexure hinge design, the most important one being the degree of freedom. For this introductory text, it may be adequate to keep the discussion confined to symmetric flexure hinges with a single degree of freedom. In the earlier days, such flexure hinges were restricted to primarily semicircular shaped notches. But subsequently, flexures with different shapes have been considered. Figure 5.9 shows some of the commonly used flexure hinge shapes.

The use of symmetric semicircular notch-type flexure hinges was popular, and more research has indicated that the performance can be superior when the notch shape is decided through an optimization process leading to smallest value of the

Table 5.2 Structural architecture of mechanisms

Type of mechanism	Architecture	Output force	Actuator location	Moving mass	Speed	Rigidity	Workspace	Structure
Stewart platform type	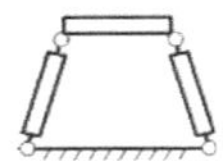	Large	On moving parts	Large	–	Good	Small	Large moving parts
Rotary actuation type	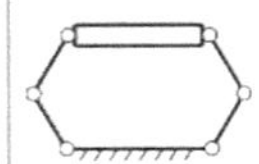	Small	Base	Small	Very fast	Poor	Large	Simple and compact
Fixed linear actuation type	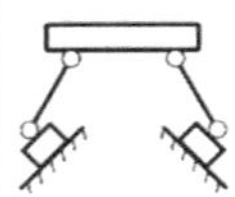	Large	Base	Small	Fast	Good	Large	Large base size

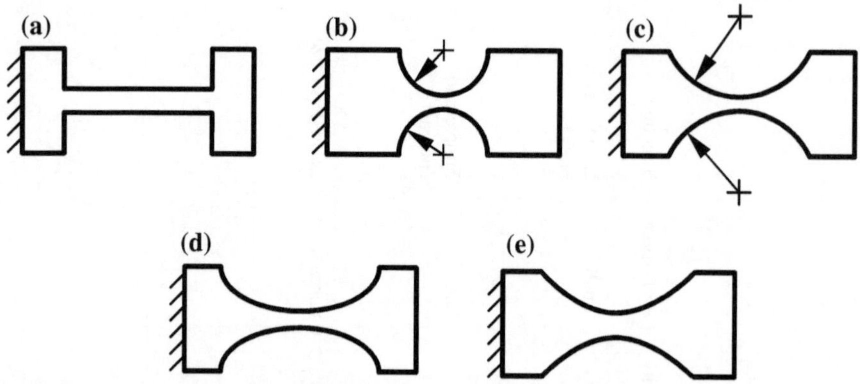

Fig. 5.9 Common flexure hinge shapes. **a** Rectangular. **b** Semicircular. **c** General circular. **d** Elliptic. **e** Polynomial curve-based flexure

parameter (stress/deflection) is often taken as a primary performance index. All the different types of flexure hinges, i.e., (a) rectangular, (b) semicircular, (c) general circular, (d) elliptic, and (e) polynomial curve types, will be discussed. Figure 5.10 shows the axes and three important types of notches.

The magnitude of stiffness for bending about the z-axis and the maximum stress developed in the material are presented below in Table 5.3.

To compare the basic performance characteristics of these three types of flexural hinges, the stiffness and stress expressions can be examined as done below in Table 5.4.

Therefore, for given E, b, h and L

$$K_{z\text{Rect}} < K_{z\text{Ellip}} < K_{z\text{Circ}}$$

and for given range of rotation θ

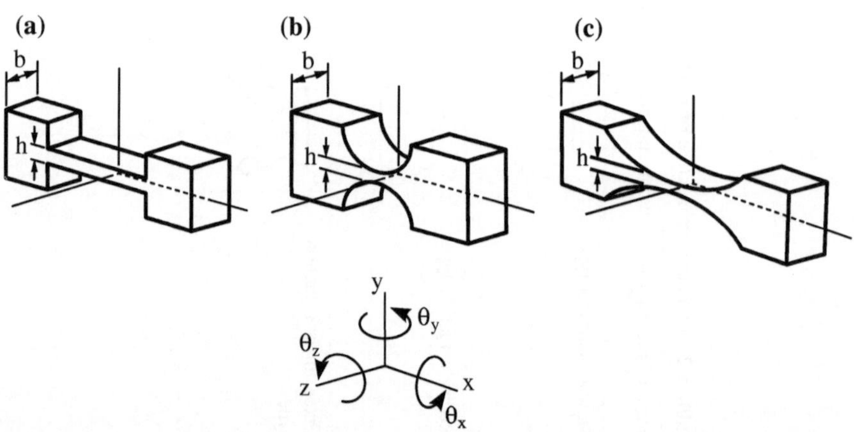

Fig. 5.10 Important types of notches for flexure hinges. **a** Rectangular. **b** Circular. **c** Elliptic

Table 5.3 Stiffness and stress developed for different notch geometries

Notch geometry	Rotational stiffness	Rotation—stress relation
Rectangular 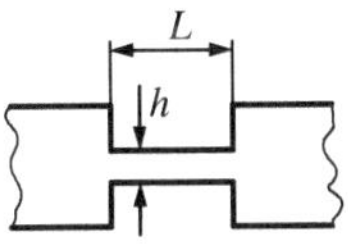	$K_z = \frac{M_z}{\theta_z} = \frac{EI_{zz}}{L} = \frac{Ebh^3}{12L}$	$\sigma_{max} = \frac{Eh}{2L}\theta_{max}$
Circular 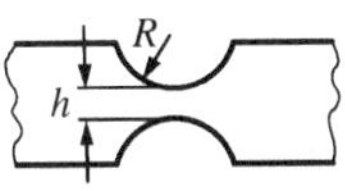	$K_z = \frac{M_z}{\theta_z} = \frac{2Ebh^{5/2}}{9\pi R^{1/2}}$	$\sigma_{max} = \frac{2\sqrt{2}E\sqrt{\frac{h}{R}}\left(1+\frac{h}{4R}\right)^{9/20}}{2L}\theta_{max}$
Elliptic 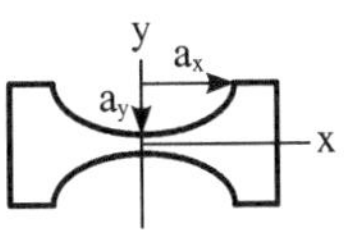	$K_z = \frac{M_z}{\theta_z} = \frac{2Eba_y^2}{3}\frac{e}{f(\beta_y)}$ $\beta_x = \frac{h}{2a_x};\ \beta_y = \frac{h}{2a_y}$ $e = \frac{a_x}{a_y} = \frac{\beta_y}{\beta_x};\ f(\beta) = \frac{3\pi}{(2\beta)^{5/2}}$ For small β $K_z = \frac{2Eba_x^2}{3e^3 f(e\beta_x)}$	

Table 5.4 Comparative performance characteristics of three types of flexure joints

Type of notch	Stiffness	Stress
Rectangular	$K_z \propto h^3/L$	$\sigma \propto h/L$
Circular	$K_z \propto h^{5/2}/L^{1/2}$	$\sigma \propto \sqrt{h/L}$
Elliptical	$K_z \propto h^{5/2}a_y^{1/2}/L$	–

$$\sigma_{Rect} < \sigma_{Ellip} < \sigma_{Circ}.$$

Thus, when higher stiffness and smaller rotation are desired, circular notch flexural hinges should be chosen. Rectangular-type leaf springs will be more suitable for low stiffness and large range of rotation.

The right circular flexure hinges are most commonly used, and therefore, the approximate expressions for some other characteristics can be sometimes useful. The torsional stiffness (rotation about the x-axis in Fig. 5.10) of this hinge can be expressed as follows:

$$\frac{M_x}{\theta_x} = \frac{8}{9\pi}Gb\frac{h^{5/2}}{\sqrt{R}} \quad (5.2.2)$$

when G is the shear modulus of the material.

The bending stiffness in the transverse direction (about y-axis) is given by

$$\frac{M_x}{\theta_x} = \frac{1}{12\pi} E b^3 \sqrt{\frac{h}{R}} \tag{5.2.3}$$

and linear stretching stiffness in the x-direction is expressed as follows:

$$\frac{F_x}{u_x} = \frac{1}{\pi} E b \sqrt{\frac{h}{R}} \tag{5.2.4}$$

where F_x and u_x are the tensile force applied in the x-direction and the corresponding stretching elongation.

It should be remembered that the above asymptotic relations are derived considering a situation in which the ratio h/R is quite small. If the ratio is higher (and even up to 0.5), the above expressions lead to erroneous results. One way to resolve the matter is to conduct an exact analysis. But, as will be shown later, the expressions are unwieldy. The other approach is to numerically compare the results from the approximate and exact analyses and derive some empirical correction factors, λ. Using this approach, the useful results can be expressed as shown below:

$$\begin{aligned}
\left(\frac{M_z}{\theta_z}\right)_{\text{corrected}} &\cong \lambda_1 \left(\frac{M_z}{\theta_z}\right); \quad \lambda_1 = 1 + 0.137\frac{h}{R} \\
\left(\frac{M_x}{\theta_x}\right)_{\text{corrected}} &\cong \lambda_1 \left(\frac{M_x}{\theta_x}\right) \\
\left(\frac{M_y}{\theta_y}\right)_{\text{corrected}} &\cong \lambda_2 \left(\frac{M_y}{\theta_y}\right); \quad \lambda_2 = 1 + 0.97\sqrt{\frac{h}{R}} \\
\left(\frac{F_x}{u_x}\right)_{\text{corrected}} &\cong \lambda_2 \left(\frac{F_x}{u_x}\right)
\end{aligned} \tag{5.2.5}$$

It has been found that using these empirical correction factors, the maximum error can be limited to about 1.5 % only.

As mentioned earlier, these flexural hinges have been subjected to considerable study and the results are well documented in the literature. Some further discussions on the topic are being presented for more involved design problems. The first major progress was made by Paros and Weisbord in 1965 and provided the oft referred results as follows:

$$\frac{\theta_z}{M_z} = \frac{3}{2EbR^2}\left(\frac{1}{2\beta+\beta^2}\right)\left\{\frac{1}{1+\beta}\cdot\frac{3+4\beta+2\beta^2}{(2\beta+\beta^2)} + \frac{6(1+\beta)}{(2\beta+\beta^2)^{3/2}}\cdot\tan^{-1}\sqrt{\frac{2+\beta}{\beta}}\right\} \tag{5.2.6}$$

where $\beta = h/2R$ and the notch is a right circular one. Further

$$\frac{\theta_y}{M_y} = 24\left(\frac{1+\beta}{\sqrt{2\beta+\beta^2}} \cdot \tan^{-1}\sqrt{\frac{2+\beta}{\beta}} - \frac{1}{\sqrt{2}}\right) \tag{5.2.7}$$

Though originally it was claimed to be an exact solution, later it has been found that expressions do not yield exact results. The reason is said to be that linear beam theory for long beams with gradually varying cross section was used by Paros and Weisbord. Finite element analysis has been used by some subsequent researchers. The important results are presented in tabular form as given in Table 5.5.

More general case of circular notch flexure hinge has been also studied. Such a flexure hinge is shown in Fig. 5.11. The left end is fixed and the right face rotations and deflections are θ_x, θ_y, θ_z, Δx, Δy and Δz.

Wu and Zhou have derived the following expressions first, using $\rho = R/h$.

$$f_1 = \frac{2(2\rho+1)}{\sqrt{4\rho+1}}\tan^{-1}\left(\sqrt{4\rho+1}\tan\frac{\phi}{2}\right) - \phi \tag{5.2.8a}$$

Table 5.5 Stiffness expressions by different authors

Author	$\frac{M_z}{\theta_z}$	$\frac{M_y}{\theta_y}$	$\frac{M_x}{\theta_x}$
Zangh and Fasse	$Ebh^2(-0.035\beta + 0.111\sqrt{\beta})$	$Eh^3(0.013\beta + 0.030\sqrt{\beta})$	$Gb^3(0.052\beta^{1.5} + 0.047\beta - 0.0082\sqrt{\beta})(0.274\gamma^2 - 1.397\beta + 2.496)$
Koster	$0.093\ Ebh^2\sqrt{\beta}$	$0.04\ Eb^3\sqrt{\beta}$	$0.0467\frac{E}{G}b\frac{\sqrt{\beta}}{1.2+\frac{1}{\beta}}$
Smith and Chetwynd	$Ebh^2\frac{\beta}{1.992+13.56\beta}$	–	–

$\beta = \frac{h}{2R}$; $\gamma = \frac{b}{2R}$

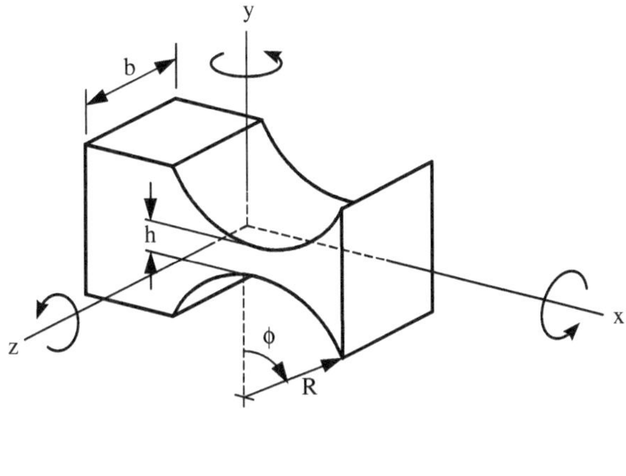

Fig. 5.11 Detailed geometry of a circular notch-based flexure hinge

$$f_2 = \frac{8\rho^4(2\rho+1)}{(4\rho+1)^2} \cdot \frac{\tan\frac{\phi}{2}}{\left[1+(4\rho+1)\tan^2\frac{\phi}{2}\right]^2} + \frac{4\rho^3(6\rho^2+3\rho+1)}{(4\rho+1)^2} \cdot \frac{\tan\frac{\phi}{2}}{1+(4\rho+1)\tan^2\frac{\phi}{2}}$$
$$+ \frac{12\rho^4(2\rho+1)}{(4\rho+1)^{5/2}} \tan^{-1}\left(\sqrt{4\rho+1}\tan\frac{\phi}{2}\right) \tag{5.2.8b}$$

$$f_3 = \frac{2\tan\frac{\phi}{2}}{\left(1+\tan^2\frac{\phi}{2}\right)^2} + \frac{\rho+1}{\rho} \cdot \frac{\tan\frac{\phi}{2}}{1+\tan^2\frac{\phi}{2}} - \frac{(2\rho+1)\sqrt{4\rho+1}}{2\rho^2} \tan^1\left(\sqrt{4\rho+1}\tan\frac{\phi}{2}\right)$$
$$+ \frac{2\rho^2+4\rho+1}{4\rho^2}\phi \tag{5.2.8c}$$

$$f_4 = \frac{2\rho^2(2\rho+1)}{(4\rho+1)} \cdot \frac{\tan\frac{\phi}{2}}{\left[1+(4\rho+1)\tan^2\frac{\phi}{2}\right]^2} + \frac{\rho(2\rho^2+5\rho+1)}{(4\rho+1)} \cdot \frac{\tan\frac{\phi}{2}}{1+(4\rho+1)\tan^2\frac{\phi}{2}}$$
$$+ \frac{(2\rho+1)(2\rho^2-4\rho-1)}{2(4\rho+1)^{3/2}} \tan^{-1}\left(\sqrt{4\rho+1}\tan\frac{\phi}{2}\right) + \frac{\rho}{4} \tag{5.2.8d}$$

Using these functions, the stiffness expressions have been derived as follows:

$$\frac{M_z}{\theta_z} = \frac{EbR^2}{12f_2} \tag{5.2.9a}$$

and the cross-stiffness for bending about z-axis is given by

$$\frac{F_y}{\theta_z} = -\frac{EbR}{12\sin\phi \cdot f_2} \tag{5.2.9b}$$

The bending about y-axis is given as follows

$$\frac{M_y}{\theta_y} = \frac{Eb^3}{12f_1} \tag{5.2.10a}$$

$$\frac{F_z}{\theta_y} = \frac{Eb^3}{12R\sin\phi \cdot f_1} \tag{5.2.10b}$$

The linear displacements can be expressed as follows:

$$\frac{M_y}{\Delta z} = \frac{Eb^3}{12R\sin\phi \cdot f_1} \tag{5.2.11a}$$

$$\frac{F_z}{\Delta z} = \left[\frac{12R^2 \sin^2 \phi}{Eb^3} f_1 + \frac{12R^2}{Eb^3} f_3\right]^{-1} \tag{5.2.11b}$$

$$\frac{M_z}{\Delta y} = -\frac{EbR}{12 \sin \phi \cdot f_2} \tag{5.2.12a}$$

$$\frac{F_y}{\Delta y} = \left[\frac{12 \sin^2 \phi}{Eb} f_2 + \frac{12}{Eb} f_4\right]^{-1} \tag{5.2.12b}$$

$$\frac{F_x}{\Delta x} = \frac{Eb}{f_1} \tag{5.2.13}$$

The above results can provide the required stiffness (or compliance values) depending on the situation specified by the value of ϕ that can lie between 0° and 90°. When ϕ = 90°, the case becomes that of a right circular flexure hinge.

The deflections, both rotary and linear, accompanying a right circular flexure joint (Fig. 5.12), depend on the forcing elements as follows:

$$\{x\} = \left[\Delta x, \Delta y, \Delta z, \theta_x, \theta_y, \theta_z\right]^{\mathrm{T}} = [C]\{F\} \tag{5.2.14}$$

where the force matrix $\{F\} = [F_x, F_y, F_z, M_x, M_y, M_z]^{\mathrm{T}}$ and $[C]$ is the compliance matrix

$$[C] = \begin{bmatrix} 0 & 0 & c_5 & 0 & 0 & 0 \\ 0 & c_3 & 0 & c_4 & 0 & 0 \\ c_1 & 0 & 0 & 0 & -c_2 & 0 \\ 0 & 0 & 0 & 0 & 0 & c_8 \\ -c_2 & 0 & 0 & 0 & c_7 & 0 \\ 0 & c_4 & 0 & c_6 & 0 & 0 \end{bmatrix}$$

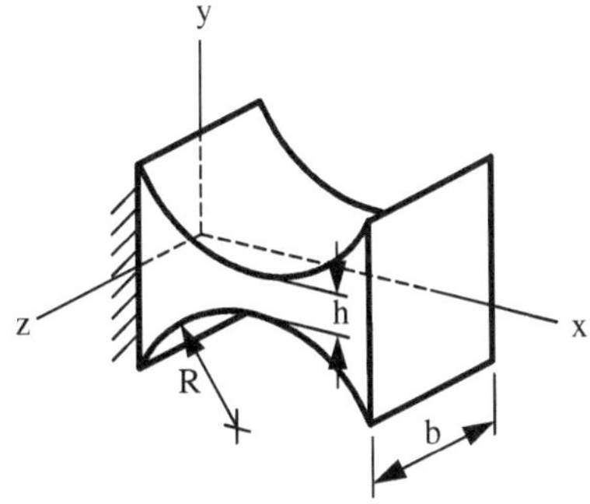

Fig. 5.12 Right circular flexure joint

The elements of the compliance matrix (for a right circular notch flexure) are given below:

$$c_1 = \frac{12\pi R^3}{Eb^3}\left\{\left(\frac{R}{h}\right)^{1/2} - \frac{1}{4}\right\}; \qquad c_2 = \frac{12R}{Eb^3}\left\{\pi\left(\frac{R}{h}\right)^{1/12} - \frac{2+\pi}{2}\right\}$$
$$c_3 = \frac{9\pi}{2Eb}\left(\frac{R}{h}\right)^{5/2} + \frac{3\pi}{2Eb}\left(\frac{R}{h}\right)^{3/2}; \qquad c_4 = \frac{9\pi R^{3/2}}{2Ebh^{5/2}}$$
$$c_5 = \frac{1}{Eb}\left\{\left(\frac{R}{h}\right)^{1/2} - \frac{\pi}{2}\right\}; \qquad c_6 = \frac{9\pi R^{1/2}}{2Ebh^{5/2}}$$
$$c_7 = \frac{12}{Eb^3}\left\{\pi\left(\frac{R}{h}\right)^{1/2} - \frac{2+\pi}{2}\right\}; \qquad c_8 = \frac{9\pi 1 R^{1/2}}{2Gbh^{5/2}}$$

During designing a complete mechanism consisting of such flexural elements, a matrix formulation can be very useful.

Though the use of flexure hinges with circular notch has been used extensively because of ease of fabrication, etc., it does not necessarily lead to the best design. The primary objective of designing a flexure hinge is to realize an optimal notch geometry that results in maximum possible allowable deflection under a specified condition. It has been found that an optimal solution can be found by using a polynomial for describing the notch contour as shown in Fig. 5.13.

The local thickness at a distance x from the origin is expressed in the form of the following polynomial function:

$$h(x) = a_0 + a_1x + a_2x^2 + \cdots + a_nx^n \tag{5.2.15}$$

with the following constraints $h(0) = h(2L) = h_{\max}$, $h(L) = h_{\min}$, $\left.\frac{\partial h}{\partial x}\right|_{x=L} = 0$.

A symmetry condition $h(L - \zeta) = h(L + \zeta)$ can be also imposed. It has been demonstrated that the use of biquadratic functions like $h(x) = 0.1 + 0.9(x - 1)^4$ in non-dimensional form as notch contours are better compared to conventional contours like circular, elliptic, etc.

So far, only the matter of compliance of a notched flexure hinge has been considered. However, to simulate a real ideal hinge, the position of the centre of rotation should also remain invariant. Thus, in designing accurate flexure hinge, kinematic constraints also need to be imposed while optimizing the notch contour.

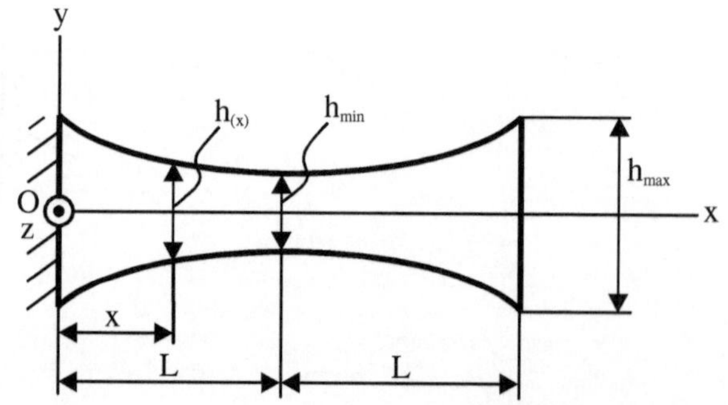

Fig. 5.13 Polynomial curve-based flexure hinge

The effect of these parasitic motions results in erroneous final required motion when the system is synthesized using the equivalent rigid body mechanism approach.

Figure 5.14a shows a right circular notch flexure hinge. In Fig. 5.14b, a quarter of the flexure hinge is shown in deflected condition, though the flexure hinge centre point A should represent the rotation point of the equivalent revolute joint at all configurations. But in reality, the instantaneous centre for the rigid part represented by the line BC shifts to position A′. A free-form notch contour can be designed through an optimization program aiming at minimizing the parasitic horizontal translation 'e'.

It can be easily recognized that 'e' depends on the degree of rotation also and is maximum when the rotation is θ_{zmax}. On the other hand, the point of intersection of line A′B with the original line AB, O does not change its position perceptibly. The position is given by the following approximate relation

$$f \approx \frac{e}{\theta_{zmax}}. \tag{5.2.16}$$

If the position of O is found out, the actual kinematic performance of the mechanism can be estimated with better precision. The optimization process for deciding the notch contour is done with an objective to minimize $1/\theta_{zmax}$ with the constraints $(\sigma - \sigma_{allow}) < 0$ and $(e - e_{imposed}) < 0$.

Another approach has been developed to account for the parasitic movements developed during elastic deformation of notched flexure hinges. Figure 5.15a shows a typical flexure hinge.

Its idealized equivalent rigid body representation is shown in Fig. 5.15b as a hinge with a torsional spring indicating the elastic nature. But it has already been shown that the relative instantaneous centre of two members 1 and 2 is not fixed as it happens in the case of a real hinge joint. There are extra 2 degrees of freedom for the parasitic movements. Thus, in total, the flexure hinge provides 3 degrees of freedom for a link when the other one is fixed (1 for rotation and other 2 for the linear displacements in the xy plane). Hence, the true modelling of the flexure hinge should

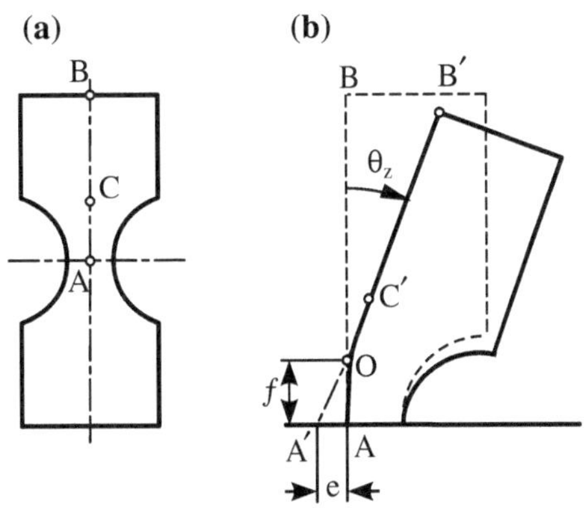

Fig. 5.14 Deformation of a right circular notch-type flexure hinge. **a** Notch geometry with reference points. **b** Deformed notch geometry

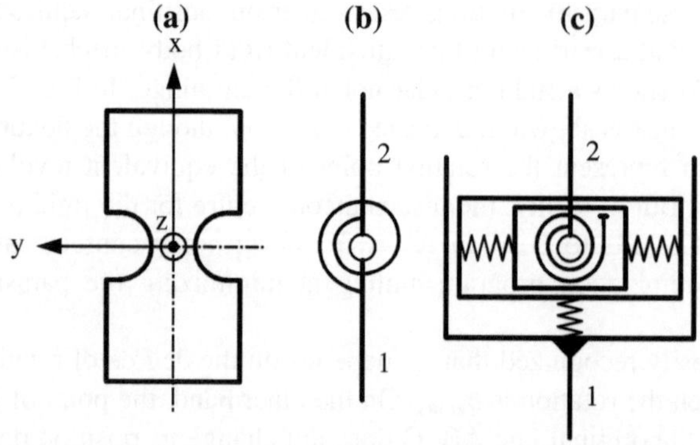

Fig. 5.15 Modelling of a flexure hinge for parasitic movements. **a** A circular notch flexure hinge. **b** Simplified equivalent revolute joint. **c** A composite joint to take care of parasitic movements

be as shown in Fig. 5.15c. Detailed kinetostatic analysis of mechanisms using 3 degree of freedom model of flexure hinges has been conducted by researchers. The characteristics of each flexure can be determined by finite element analysis.

Different combinations of simple notch-type flexure hinges give rise to different types of relative motion characteristics of the connected links. Figure 5.16 shows how a combination of flexure hinges results in an equivalent prismatic pair. When a force F is applied to the moving member 2, it executes a linear translation by amount Δ as shown. The bottom pair of flexures support the total moment FD. If the rotation of each joint be θ_z, then the displacement Δ will be $\theta_z D$.

Hence, the resultant stiffness of this prismatic joint will be

$$k = \frac{F}{\Delta}$$

Using approximate relations developed by Paros and Weisbord, the final expression for the stiffness becomes as follows:

$$k \cong \frac{8Ebh^{5/2}}{9\pi D^2 R^{1/2}} \tag{5.2.17}$$

Fig. 5.16 Combination of flexure hinges to result in an equivalent prismatic joint

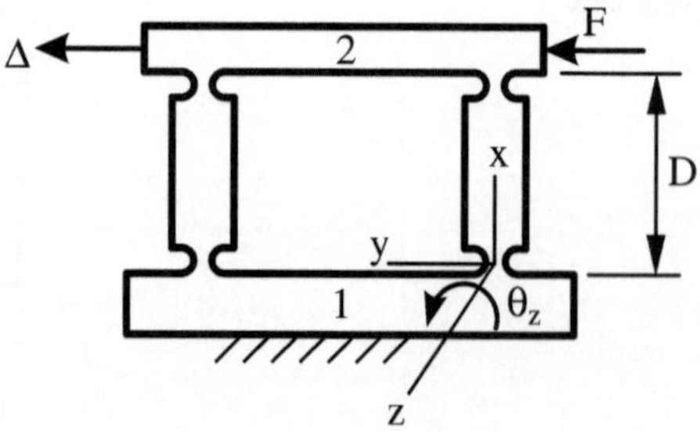

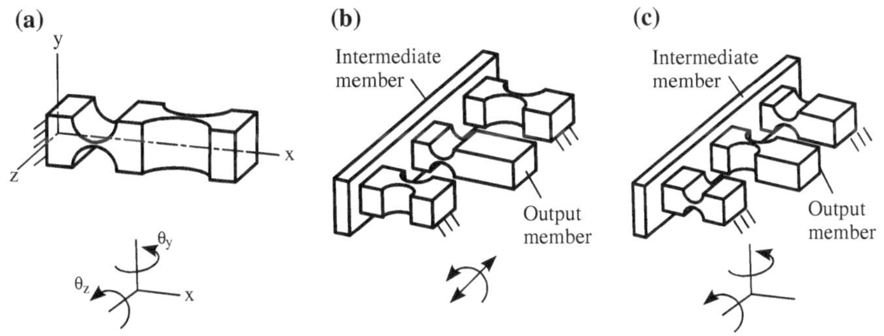

Fig. 5.17 Composite joint structures for more than one degree of freedom. **a** Two rotational degrees of freedom. **b** One translational and one rotational degrees of freedom. **c** Two rotational degrees of freedom

For thicker notches (5.2.5) can be used along with the correction factor λ_1, and the equivalent stiffness can be expressed as follows:

$$k \approx \frac{Ebh^3}{6\lambda_1 RD^2} \tag{5.2.18}$$

Composite flexure notches can result in equivalent hinges with higher degrees of freedom. Figure 5.17a–c shows the formation of flexure hinges with two degrees of freedom and can represent conventional universal joints.

The composite structure shown in Fig. 5.17a and c provides two rotational degrees of freedom about two mutually perpendicular axes. On the other hand, the flexure composite joint shown in the Fig. 5.17b provides translational and a rotational degree of freedom. Another type of flexure hinge, called spherical joint, providing three rotational degrees of freedom, is shown in Fig. 5.18.

It is not frequently used because of fabricating such miniaturized notches. Paros and Weisbord have shown that the angular rotations θ_y and θ_z due to an applied moment M can be approximately expressed according to the following relation:

$$\theta_y = \theta_z \approx \frac{20M\sqrt{R}}{Eh^{7/2}} \tag{5.2.19}$$

where h is the smallest diameter of the notch and R is the notch radius (Fig. 5.18b).

5.2.1.2 Synthesis of Compliant Micromechanisms Using Flexure Hinges

The simplest approach for synthesizing micromechanisms is to synthesize first the equivalent rigid body mechanism (i.e. mechanism with rigid bodies connected by joints to allow relative motion) for generating the desired motions. Then, the joints

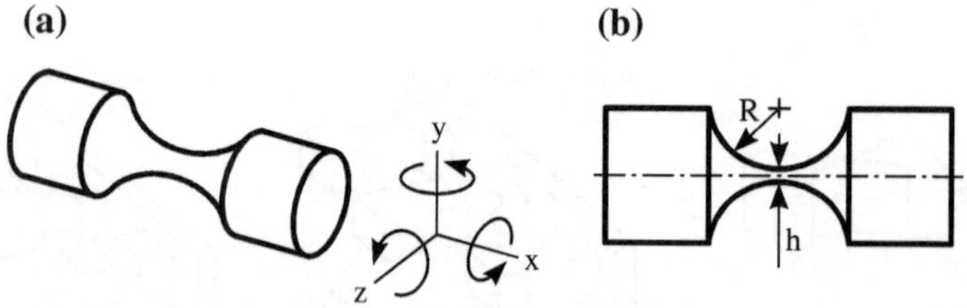

Fig. 5.18 Spherical flexure hinge. **a** Basic shape and rotational degrees of freedom. **b** Geometry of a spherical flexure hinge

are replaced by suitable flexure hinges described earlier in this section. However, since the flexure hinges do not represent the conventional hinges exactly, this rigid body model with flexures does not produce the desired motion correctly. The relative pure rotation between two rigid bodies connected by a hinge is not exactly reproduced when the hinge is replaced by a flexure. There can be three options. If the accuracy requirement is not that stringent, a simple rigid body model (sometimes called equivalent rigid body model or ERB model) can lead to satisfactory results. If the flexure hinges are specially designed with optimized contours to keep the parasitic motions minimal, then ERB with such optimally designed flexures can lead to reasonably accurate results. To achieve still better results, the mechanism obtained by an equivalent rigid body model is used as a first approximation. Then, the whole mechanism is modified through successive iterations till an optimal solution is achieved. It is not necessary to use flexures with special notch contour geometry; instead, standard flexures with circular notches can be used. The successive modifications gradually eliminate the errors due to parasitic motions in the flexure hinges. The effect of parasitic motions in the flexures can be incorporated in many ways. But the most direct approach is to modify the equivalent link lengths continuously as a function of angular motions of the flexures. Figure 5.19 shows a kinematic chain where the rigid links are connected by flexure hinges. In the situation when the flexure hinges are in undeformed condition, the nominal lengths of equivalent links can be determined by joining flexure centre points (as shown in Fig. 5.19 as A, B, C, D) as l_{i-1_0}, $l_{i_0,i+1_0}$.

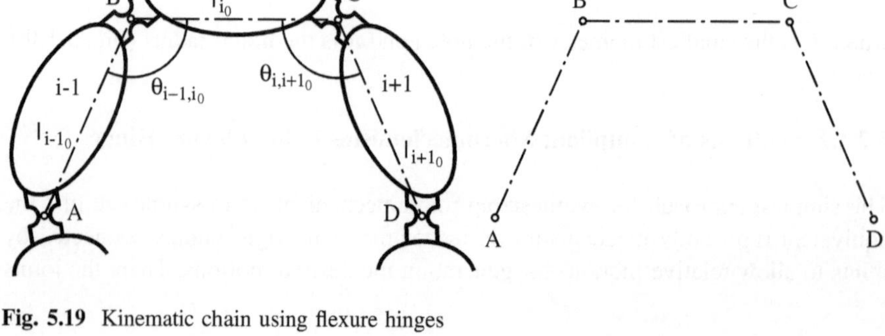

Fig. 5.19 Kinematic chain using flexure hinges

As the instantaneous centre for rotation shifts from the mid-point of a flexure as discussed in the earlier section, the equivalent link lengths also change. Thus, these equivalent link lengths become functions of the flexure bending angles. If the angle between links $i-1$ and i in the undeformed state of a compliant mechanism be θ_{i-l,i_0}, then any change in this represents the amount of bending of the flexure hinge connecting these two rigid links. So, the instantaneous equivalent link length of link i depends upon the bending of the flexure hinges connecting link i to its neighbouring links say, $i-1$ and $i+1$. If the angle subtended by two successive equivalent links $i-1$ and i be $\theta_{i-1,i}$, then

$$l_i = l_i\left(l_{i_0}, \theta_{i-1,i_0}, \theta_{i,i+1}\right) \tag{5.2.20}$$

Hence, the small change in this length at a different configuration of the mechanism can be expressed as follows:

$$\delta l_i = \frac{\partial l_i}{\partial \theta_{i-1,i}} \delta \theta_{i-1,i} + \frac{\partial l_i}{\partial \theta_{i,i+1}} \delta \theta_{i,i+1}. \tag{5.2.21}$$

where $\delta\theta_{i-1,i}$ represents the change in the angle $\theta_{i-1,i}$ from its original undeformed condition. Dividing the complete cycle of motion into small steps, the position of the output member can be found out by continuously revising the values of the equivalent rigid body lengths. If the error exceeds the acceptable limits, the value of l_{i_0}, obtained by the initial equivalent rigid body mechanism, synthesized following the standard approaches for mechanism synthesis, should be modified. The process is repeated methodically till the errors in the generated motion are within acceptable limits.

5.2.2 Micromechanisms with Distributed Compliance

Considering the primary objectives of a mechanism—(i) maximizing the flexibility to achieve desired output motion and (ii) maximizing the overall stiffness of the structure for providing maximum load bearing capacity—it is obvious that the best solution is the rigid link mechanism with hinge joints. Any computational technique to generate compliant topologies that imitate such rigid link mechanisms leads to mechanisms with lumped compliance-based mechanisms described in the previous section. Thus, flexure hinge-based pseudo-rigid-body models are nearest to the ideal situation. But such flexure hinge-based mechanisms are not useful when large angular displacements and high strength are desired. The primary source of weakness is of course the flexure hinges. A better solution can be obtained when the flexibility of the compliant mechanism is distributed uniformly throughout the structure instead of being at a few points only. Recent developments in materials engineering have resulted in materials which can sustain considerable deformation with adequate strength.

Work on the procedure for designing monolithic continua to undergo deformation in a desired way under applied loads (or input deformation) has been done by many researchers since the early publication by Frecker, Sigmund, Ananthasuresh, Saxena and others since 1997. In the procedure, usually termed as 'topology optimization', the whole design region of the material body is divided into finite elements in a suitable way depending on the geometry. A variable is assigned to every point in this region that is varied smoothly between 'existence' and 'non-existence' states. The example shown in Fig. 5.20 has been used by Anathasuresh et al. to demonstrate the different routes to the same final solution. Applying topology optimization directly to the design region leads to a system as shown in Fig. 5.20b.

Again, the equivalent rigid body approach leads to a system like the one shown in Fig. 5.20c, which, in essence, is the same as the solution obtained by topology optimization, as indicated in Fig. 5.20d. Thus, often a direct application of the topology optimization approach leads to lumped compliance-based system for the

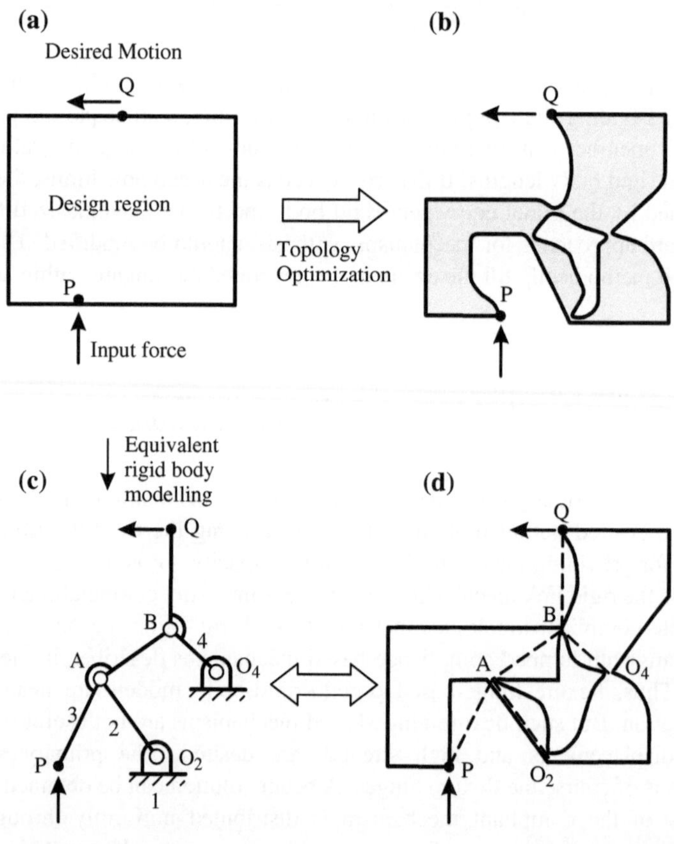

Fig. 5.20 Different approaches for compliant mechanism synthesis. **a** Basic requirement. **b** Result from direct topology optimization. **c** Equivalent rigid body mechanism model. **d** Result from equivalent rigid body modelling

obvious reason that a mechanism based on rigid links connected by hinges is the ideal solution as already stated before. These approaches based upon force–deflection-type formulations for topology optimization simultaneously maximize the deflection at the output point and maximize the overall structural stiffness. The stiffness is measured by the stain energy stored in the deformed structure. Since revolute joint-based rigid link mechanisms can produce large deformation without any strain energy stored in the system that is the ideal solution obtained by the direct straightforward applications of topology optimization approach.

As indicated earlier many times, the monolithic mechanisms with lumped compliances (in the form of flexure hinges) are not suitable. Compliant mechanisms with distributed compliance are useful in such situations. Intuitively, it appears that such systems can be obtained by following a route in which the stress (or strain energy) is more or less uniformly distributed throughout the whole structure. Unfortunately, this is not a very helpful technique for avoiding the occurrence of flexure hinges (lumped compliances). This is because of the fact that the strain energy in flexure hinges is not any larger than at other elements in the system. To avoid the occurrence of lumped compliances, a method has been developed by Anathasuresh and others in which occurrence of point flexures is heavily penalized.

Chapter 6
Dynamics of Micromechanisms

Abstract The dynamical problems for systems at microscale have different characteristics. Both free oscillation and forced oscillation of micromechanisms with flexure hinges are discussed. The concept of shift and spread during forced oscillation is explained. The dynamical analysis of mechanisms with flexible links is presented at the end. This approach is necessary for understanding the dynamics of micromechanisms with distributed compliance.

Keywords Dynamics of micromechanisms · Forced and free oscillation · Dynamics of micromechanisms with distributed compliance · Spread and shift

6.1 Dynamical Problems at Small-Scale Systems

The problem of dynamics of micromechanisms can be grouped under two separate classes. Those types of micromechanisms which are assembled from individual microcomponent possess no self-oscillating characteristics unless elastic elements are imposed upon these systems. The other group consists of problems involving monolithic mechanisms with lumped and distributed compliance. In both cases, micromechanisms possess characteristic equilibrium position and a characteristic frequency of natural oscillation about the equilibrium position when distributed. Thus, the assembled micromechanisms need not possess any ability to execute free oscillation, and therefore, no resonant situation can arise when subjected to external forces—either static or dynamic. However, in most cases in practice, the micromechanisms consist of rigid links connected through elastic joints and their motion under an external dynamic excitation needs careful analysis. The problem of determining the response under an external force (or moment) has a few aims. If the externally applied disturbance is a static force (or moment), it is important to find out the corresponding equilibrium configuration of the mechanism. On the other hand, if the disturbance is a dynamic force (or moment), the equilibrium position

A. Ghosh and B. Corves, *Introduction to Micromechanisms and Microactuators*,
Mechanisms and Machine Science 28, DOI 10.1007/978-81-322-2144-9_6

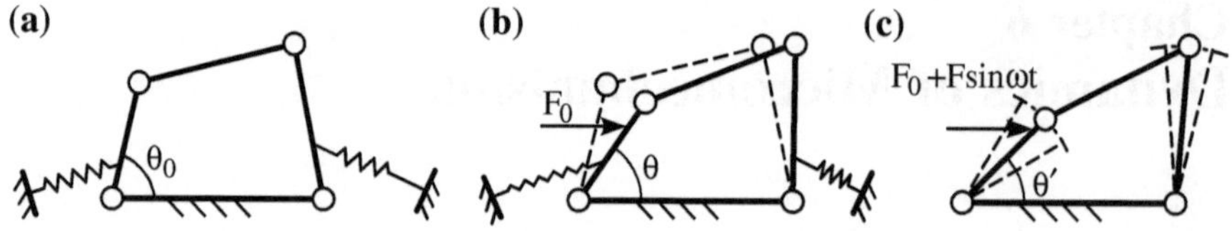

Fig. 6.1 Dynamics of mechanisms with elastic connection. **a** A four-bar mechanism with elastic connections in equilibrium position. **b** Response to a static force. **c** Response to a dynamic force

will be different. The amplitude will depend on the excitation and system parameters. The matter is explained in Fig. 6.1.

The four-bar mechanism is constrained by connecting its links to some elastic connections, and it possesses an equilibrium position as indicated by the input member angle θ_0, shown in Fig. 6.1a. When a static force F_0 is applied, the mechanism takes a different position corresponding to an input link angle θ as shown in Fig. 6.1b. When the applied force is a dynamic quantity $F_0 + F\sin\omega t$, the mechanism oscillates about a different equilibrium position given by angle θ' of the input link even though the mean force is F_0. This is shown in Fig. 6.1c. Thus, the equilibrium positions under a given force can be different depending on the dynamic nature of the given force, the mean value remaining the same. This counter-intuitive phenomenon makes a dynamic analysis of compliant mechanisms very essential.

Micromechanisms with elastic members also possess a capability to execute free oscillation about an equilibrium position. Dynamic analysis is, of course, more involved and is generally accomplished through the use of computational mechanics techniques such as finite element method. As seen from Table 5.4, the stiffness of flexure hinges is proportional to h^3/L, $h^{5/2}/L$ and $h^{5/2}a_y^{1/2}/L$, depending on the shape of the notch. Thus, in all such cases, the stiffness scales as l^2. On the other hand, the mass, which is proportional to volume for a given material, scales as l^3. Hence, the natural frequency scales as l^{-1} and, as a consequence, is very high in case of micromechanisms. Therefore, the characteristic frequency of the applied force (or moment) is much lower than the natural frequency of the mechanism, and a kinetostatic analysis is good enough, the inertia being very low. But for high frequency disturbances, a dynamic analysis is desirable. But the effects of friction and other adhesive phenomena become relatively very important in case of multimember assembled micromechanisms. Because of the relatively higher importance, the dynamics of compliant micromechanisms will be taken up first for discussion.

6.2 Dynamics of Compliant Micromechanisms with Flexure Hinges

Compliant micromechanisms with flexure hinges will be considered first for dynamic modelling. Such systems are not necessarily connected to external elastic members but because of their monolithic structures with some flexure hinges

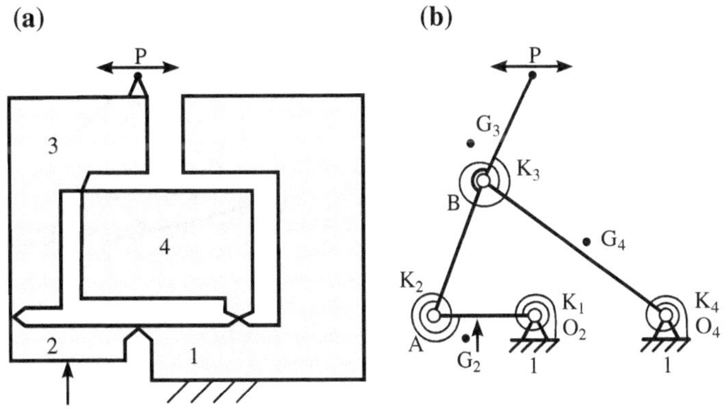

Fig. 6.2 Chebyshev's straight line micromechanism. **a** Micromechanism with flexure hinges. **b** Equivalent mechanism with revolute joints

provide the required elastic character represented by an equilibrium position and a natural frequency. Not to make analysis too involved, the flexure hinges will be considered as simple hinges connected to torsional springs as shown in Fig. 6.2b. But for a more accurate analysis, each flexure hinge needs to be provided with three degrees of freedom (to take care of the parasitic motions) as indicated in Fig. 5.15c.

As an example, let the Chebyshev's straight line mechanism, shown in Fig. 6.2a, b, be considered. The micromechanism with elastic hinges is shown in Fig. 6.2a, and the equivalent linkage is shown in Fig. 6.2b. The centre of mass of the three moving links 2, 3 and 4 is at locations G_2, G_3 and G_4, respectively, as shown in Fig. 6.2b. The corresponding mass moments of inertia about axes passing through these points are J_2, J_3 and J_4 respectively, the masses of the moving links being m_2, m_3 and m_4. The four flexure hinges are replaced by four revolute joints and four torsional springs of stiffness K_1, K_2, K_3 and K_4 as shown in the figure. If the gravitational potential energy is ignored, the total potential energy of the system is confined to the strain energy of the four springs (elastic elements in the system). So,

$$U = \frac{1}{2}\sum_{i=1}^{4} K_i\phi_i^2 \tag{6.2.1}$$

where ϕ_i represents the angular deformation of the ith torsional spring. It is also obvious that the equilibrium position of the system will correspond to the situation when U attains the minimum value. Now, for a four-bar mechanism, there is one degree of freedom.

If 2 is chosen as the input link, quite often the angle θ_2, which this makes with a reference line, say O_2O_4 (Fig. 6.3), is chosen as the input parameter that specifies the mechanism's configuration. So, once θ_2 is given, the mechanism configuration is known and the values of ϕ_i (i = 1, 2, 3, 4) are determined. Hence, U can be considered as a function of θ_2, i.e.

Fig. 6.3 Input parameter of a four-bar mechanism

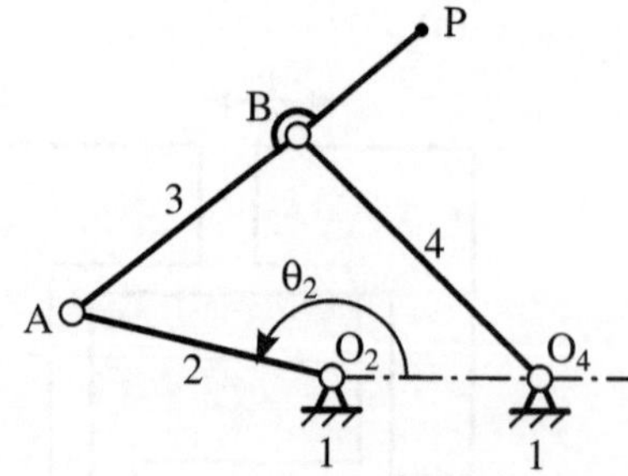

$$U = U(\theta_2) \tag{6.2.2}$$

Therefore, the equilibrium configuration corresponds to $\theta_2 = \theta_{2_0}$ so that

$$\left.\frac{\partial U}{\partial \theta_2}\right|_{\theta_2=\theta_{2_0}} = 0. \tag{6.2.3}$$

When a force $\bar{F}$ or moment M acts on the input link 2, the mechanism attains a new equilibrium position θ_2 and the work done by the externally applied force $\bar{F}$ or moment M equals the extra strain energy in the springs. The following equation is, thus, obtained

$$\bar{F} \cdot \Delta \bar{s} = W = \Delta U = U(\theta_2) - U(\theta_{2_0}) \tag{6.2.4a}$$

or

$$M \cdot (\theta_2 - \theta_{2_0}) = W = \Delta U = U(\theta_2) - U(\theta_{2_0}) \tag{6.2.4b}$$

If the minimum potential energy at $\theta_2 = \theta_{2_0}$ is set to zero as reference, then

$$W = U(\theta_2). \tag{6.2.4c}$$

So, when a force (or moment) is slowly applied, the mechanism attains a new equilibrium position that can be found out from (6.2.4c) and (6.2.4a) (or 6.2.4b). If a number of external forces and moments are applied, the work done W has to be correctly found out and this represents the potential energy of the systems.

6.2.1 Free Oscillation

As mentioned earlier, a mechanism with elastic connections (representing the flexure hinges) executes natural oscillation about its equilibrium position, when subjected to an initial disturbance. If θ_{2_0} be the angle of the input link corresponding to the equilibrium position, then this free oscillation takes place about this value of the input link (or any reference link for a single degree of freedom system). To derive the equation of motion for free oscillation, the most generalized approach is to use Lagrange's equation. For constrained mechanisms (i.e. mechanisms with single degree of freedom), the deviation from the equilibrium position (represented by θ_{2_0}) can be specified by an angle ψ $(=\theta_2 - \theta_{2_0})$ which is small for small oscillations. If ψ is taken as the generalized coordinate, U and T be the instantaneous potential and kinetic energies, respectively, of the system, then the Lagrange's equation of motion becomes

$$\frac{\mathrm{d}}{\mathrm{d}t}\left(\frac{\partial T}{\partial \dot{\psi}}\right) - \frac{\partial T}{\partial \psi} = -\frac{\partial U}{\partial \psi} \tag{6.2.5}$$

where $T = T\left(\psi, \dot{\psi}\right)$ and $U = U(\psi)$ and equilibrium position corresponds to $\psi = 0$. Let G_2, G_3 and G_4 be the locations of the centres of the mass for the moving links and let $\bar{v}_{G_2}$, $\bar{v}_{G_3}$ and $\bar{v}_{G_4}$ be their instantaneous velocities, respectively. Again, let ω_2, ω_3 and ω_4 be the instantaneous angular velocities of the moving links. Then,

$$T = \frac{1}{2}\sum_{i=2}^{4} m_i v_{G_i}^2 + \frac{1}{2}\sum_{i=1}^{4} J_{G_i}\omega_i^2 \tag{6.2.6}$$

where m_i and J_{G_i} be the mass and moment of inertia about axes passing through the centres of mass, respectively. In case of constrained mechanisms, v_{G_i} and ω_i can be expressed in terms of $\dot{\psi}$ and ψ.

So far as the deformation of the elastic elements, ϕ_i, is concerned, they can be represented as a function of $\theta_{2_0} + \psi$. Thus,

$$U = \frac{1}{2}\sum_{i=1}^{4} K_i\phi_i^2 = \frac{1}{2}\sum_{i=1}^{4} K_i\phi_i^2(\theta_{2_o} + \psi_i) \tag{6.2.7}$$

and $U = U(\psi)$. Now, by Taylor series expansion,

$$\frac{\partial U}{\partial \psi} = \left(\frac{\partial U}{\partial \psi}\right)_{\psi=0} + \left(\frac{\partial^2 U}{\partial \psi^2}\right)_{\psi=0} \cdot \psi + \cdots.$$

Since $\psi = 0$ implies the equilibrium position where U is minimum $(\partial U/\partial\psi)_{\psi=0} = 0$. Hence,

$$\frac{\partial U}{\partial \psi} \cong \left(\frac{\partial^2 U}{\partial \psi^2}\right)_{\psi=0} \cdot \psi \approx K_{\mathrm{eq}0}\psi \tag{6.2.8}$$

where $K_{\mathrm{eq}0}$ can be considered to be the equivalent torsional spring stiffness with input link as reference at the equilibrium position. Again from (6.2.6),

$$T \approx \frac{1}{2}\dot{\psi}^2 \sum_{i=2}^{4} \left[m_i \left(\frac{\mathrm{d}S_{\mathrm{G}_i}}{\mathrm{d}\psi} \right)_0^2 + J_{\mathrm{G}_i} \left(\frac{\mathrm{d}\theta_i}{\mathrm{d}\psi} \right)_0^2 \right] \tag{6.2.9}$$

when the higher order terms in ψ are neglected as ψ is small, and S_{G_i} represents the position centre of mass G_i. This is so because

$$v_{\mathrm{G}_i} = \frac{\mathrm{d}S_{\mathrm{G}_i}}{\mathrm{d}\psi} \cdot \dot{\psi} \quad \text{and} \quad \omega_i = \frac{\mathrm{d}\omega_i}{\mathrm{d}\psi} \cdot \dot{\psi}.$$

From (6.2.9), the kinetic energy can be expressed as follows:

$$T = \frac{1}{2} J_{\mathrm{eq}0} \dot{\psi}^2 \tag{6.2.10}$$

where $J_{\mathrm{eq}0}$ is the equivalent moment of inertia of the system as referred to the input link at its equilibrium position. Using (6.2.5), (6.2.8) and (6.2.10), the equation of motion for free oscillation of the mechanism becomes

$$\frac{\mathrm{d}}{\mathrm{d}t}\left(J_{\mathrm{eq}0} \dot{\psi} \right) + K_{\mathrm{eq}0} \psi \approx 0$$

or

$$J_{\mathrm{eq}0} \ddot{\psi} + K_{\mathrm{eq}0} \psi = 0, \tag{6.2.11}$$

and the frequency of natural oscillation v_{n} is found out as

$$v_{\mathrm{n}} = \frac{1}{2\pi} \sqrt{\frac{K_{\mathrm{eq}0}}{J_{\mathrm{eq}0}}}. \tag{6.2.12}$$

6.2.2 Forced Oscillation

The example of the previous section will be used for explaining the dynamics of forced oscillation also, and the mechanism (equivalent rigid body model with elastic springs at the hinges) is shown once more in Fig. 6.4.

Some important matter is required to be emphasized before the analysis is presented. The equilibrium position of the mechanism, under the influence of the torsional springs representing the respective flexure hinges, is represented by the configuration $\mathrm{O_2ABO_4}$. Let the mechanism be driven by a force F applied to the input link at Q, and let the path of the point of application of the force be along F, perpendicular to $\mathrm{O_2Q}$. The small displacement of Q along F may be taken as s as indicated in the figure. When a static force F_0 is applied to the reference/input link

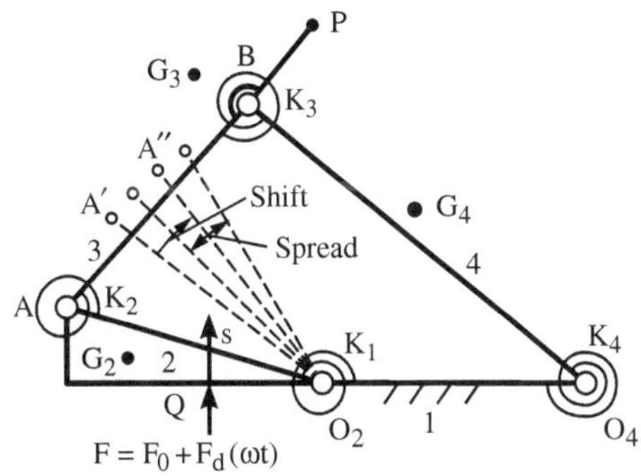

Fig. 6.4 Shift and spread of an oscillating mechanism

2, its position shifts to O_2A' as shown in the figure. However, when a dynamic force with a mean value of F_0, $F_0 + F_d(\omega t)$ is applied to the reference link 2, it does not oscillate about its equilibrium position under the influence of the mean force F_0 applied statically. Instead, the reference link oscillates about a new mean position indicated by O_2A''! The change in position from O_2A' to O_2A'' is called the 'shift', and the range of oscillation is called the 'spread' as indicated in Fig. 6.4. This phenomenon is somewhat counter-intuitive and needs to be noted carefully. For very high-precision compliant mechanisms, a complete dynamic analysis is, therefore, very important.

Thus, under the action of a periodic excitation, the angle of the reference/input link at any instant can be expressed as follows:

$$\theta_2 = \theta_{2_{\text{dyn}}} + \psi \tag{6.2.13}$$

where $\theta_{2_{\text{dyn}}}$ is the angle of the link about which the forced oscillation occurs. Or, it can be considered to be the dynamic equilibrium position of the reference link under influence of an excitation

$$F = F_0 + F_d(\omega t) \tag{6.2.14}$$

where F_d is the amplitude of the dynamic component of the external force, and ω is the circular frequency of the dynamic excitation. So,

$$\underset{t\to\infty}{\delta t}\, \frac{1}{t}\int_0^t \theta_2 \mathrm{d}t = \theta_{\text{dyn}}$$

and

$$\underset{t\to\infty}{\delta t}\, \frac{1}{t}\int_0^t \psi \mathrm{d}t = 0.$$

Since the dynamic part of the excitation is a periodic force with zero mean value

$$\frac{1}{\tau}\int_0^{\tau} F_{\mathrm{d}}(\omega t)\mathrm{d}t = 0.$$

Taking link 2 as the reference link and θ_2 as the generalized coordinate for the single degree of freedom constrained mechanism, the generalized moment acting on link 2 (in general, as a resultant effect of all the external moments and forces acting on the system with frequency ω) can be written as

$$\begin{aligned} M &= M(\theta_{2_{\mathrm{dyn}}} + \psi) \\ &= [F_0 + F_{\mathrm{d}}(\omega t)]\cdot\frac{\mathrm{d}s}{\mathrm{d}\theta_2} - \frac{\mathrm{d}U(\theta_2)}{\mathrm{d}\theta_2} \end{aligned} \tag{6.2.15}$$

As ψ is small by Taylor series expansion and neglect of the terms containing higher order of ψ the following relation is obtained

$$\begin{aligned} M &\approx [F_0 + F_{\mathrm{d}}(\omega t)]\left[\left(\frac{\mathrm{d}s}{\mathrm{d}\theta_2}\right)_{\mathrm{dyn}} + \left(\frac{\mathrm{d}^2 s}{\mathrm{d}\theta_2^2}\right)_{\mathrm{dyn}}\cdot\psi\right] - \left(\frac{\mathrm{d}U(\theta_2)}{\mathrm{d}\theta_2}\right)_{\mathrm{dyn}} - \left(\frac{\mathrm{d}^2 U}{\mathrm{d}\theta_2^2}\right)_{\mathrm{dyn}}\cdot\psi \\ &\approx F_0\left(\frac{\mathrm{d}s}{\mathrm{d}\theta_2}\right)_{\mathrm{dyn}} - \left(\frac{\mathrm{d}U}{\mathrm{d}\theta_2}\right)_{\mathrm{dyn}} + \left(\frac{\mathrm{d}s}{\mathrm{d}\theta_2}\right)_{\mathrm{dyn}}\cdot F_{\mathrm{d}}(\omega t) \\ &\quad + \left\{[F_0 + F_{\mathrm{d}}(\omega t)]\left(\frac{\mathrm{d}^2 s}{\mathrm{d}\theta_2^2}\right)_{\mathrm{dyn}}\cdot\psi - \frac{\mathrm{d}^2 U}{\mathrm{d}\theta_2^2}\cdot\psi\right\} \end{aligned} \tag{6.2.16}$$

It should be noted that the term

$$F_0\left(\frac{\mathrm{d}s}{\mathrm{d}\theta_2}\right)_{\mathrm{dyn}} - \left(\frac{\mathrm{d}U}{\mathrm{d}\theta_2}\right)_{\mathrm{dyn}} = \lambda$$

is that part of the generalized moment that arises because of the difference in θ_{2_0} and $\theta_{2_{\mathrm{dyn}}}$, the angles for the static and the dynamic equilibrium positions. Writing

$$F_0\left(\frac{\mathrm{d}^2 s}{\mathrm{d}\theta_2^2}\right)_{\mathrm{dyn}} - \frac{\mathrm{d}^2 U}{\mathrm{d}\theta_2^2} = -K_{\mathrm{dyn}}$$

Equation (6.2.16) takes the following form:

$$M(\psi) \approx \lambda + \left(\frac{\mathrm{d}s}{\mathrm{d}\theta_2}\right)_{\mathrm{dyn}} \cdot F_{\mathrm{d}}(\omega t) - \left[K_{\mathrm{dyn}} - \left(\frac{\mathrm{d}^2 s}{\mathrm{d}\theta_2^2}\right)_{\mathrm{dyn}} \cdot F_{\mathrm{d}}(\omega t)\right]\psi. \tag{6.2.17}$$

As $\theta_{2_{\mathrm{dyn}}}$ position is still an unknown, all quantities with a suffix 'dyn' represent undetermined quantities at the dynamic equilibrium position.

Now, using (6.2.10) for stating equilibrium position modified for dynamic equilibrium position

$$T \approx \frac{1}{2} J_{\mathrm{dyn}}\, \dot{\psi}^2$$

where J_{dyn} is the equivalent moment of the inertia at the reference link (representing the inertia of the whole system) for the dynamic equilibrium position (still unknown). The Lagrange equation then yields the following:

$$\frac{\mathrm{d}}{\mathrm{d}t}\left(\frac{\partial T}{\partial \dot{\psi}}\right) - \frac{\partial T}{\partial \psi} = J_{\mathrm{dyn}}\, \ddot{\psi} \tag{6.2.18}$$

J_{dyn} can be expressed as follows:

$$J_{\mathrm{dyn}} = \sum_{i=2}^{4}\left[m_i \left(\frac{\mathrm{d}s_{G_i}}{\mathrm{d}\theta_2}\right)^2_{\mathrm{dyn}} + J_i \left(\frac{\mathrm{d}\theta_i}{\mathrm{d}\theta_2}\right)^2_{\mathrm{dyn}}\right] \tag{6.2.19}$$

As

$$\frac{\mathrm{d}}{\mathrm{d}t}\left(\frac{\partial T}{\partial \dot{\psi}}\right) - \frac{\partial T}{\partial \psi} = M(\psi)$$

using (6.2.7) and (6.2.18), one gets

$$J_{\mathrm{dyn}}\, \ddot{\psi} = \lambda + \left(\frac{\mathrm{d}s}{\mathrm{d}\theta_2}\right)_{\mathrm{dyn}} \cdot F_{\mathrm{d}}(\omega t) - \left[K_{\mathrm{dyn}} - \left(\frac{\mathrm{d}^2 s}{\mathrm{d}\theta_2^2}\right)_{\mathrm{dyn}} \cdot F_{\mathrm{d}}(\omega t)\right]\psi$$

or

$$\ddot{\psi} = \frac{\lambda}{J_{\mathrm{dyn}}} + \frac{1}{J_{\mathrm{dyn}}}\left(\frac{\mathrm{d}s}{\mathrm{d}\theta_2}\right)_{\mathrm{dyn}} \cdot F_{\mathrm{d}}(\omega t) - \frac{K_{\mathrm{dyn}}}{J_{\mathrm{dyn}}} + \frac{1}{J_{\mathrm{dyn}}}\left(\frac{\mathrm{d}^2 s}{\mathrm{d}\theta_2^2}\right)_{\mathrm{dyn}} \cdot F_{\mathrm{d}}(\omega t)$$

or

$$\ddot{\psi} + \left[\omega_{n\mathrm{dyn}}^2 - \frac{1}{J_{\mathrm{dyn}}}\left(\frac{\mathrm{d}^2 s}{\mathrm{d}\theta_2^2}\right)_{\mathrm{dyn}} \cdot F_{\mathrm{d}}(\omega t)\right]\psi = \frac{\lambda}{J_{\mathrm{dyn}}} + \frac{1}{J_{\mathrm{dyn}}}\left(\frac{\mathrm{d}s}{\mathrm{d}\theta_2}\right)_{\mathrm{dyn}} \cdot F_{\mathrm{d}}(\omega t)$$

where

$$\omega_{n\mathrm{dyn}}^2 = K_{\mathrm{dyn}}/J_{\mathrm{dyn}} \tag{6.2.20}$$

When the excitation is harmonic $F_{\mathrm{d}}(\omega t) = F_0 \cos\omega t$, the analysis can be carried forward. In this case, (6.2.20) takes the following form

$$\ddot{\psi} + \left[\omega_{n\mathrm{dyn}}^2 - \frac{F_{\mathrm{d}}}{J_{\mathrm{dyn}}}\left(\frac{\mathrm{d}^2 s}{\mathrm{d}\theta_2^2}\right)_{\mathrm{dyn}} \cos\omega t\right]\psi = \frac{\lambda}{J_{\mathrm{dyn}}} + \frac{F_0}{J_{\mathrm{dyn}}}\left(\frac{\mathrm{d}s}{\mathrm{d}\theta_2}\right)_{\mathrm{dyn}} \cos\omega t. \tag{6.2.21}$$

Denoting ωt by $2z$ and using the following abbreviations

$$d = \frac{\lambda}{J_{\mathrm{dyn}}}, a = \left(\frac{\omega_{n\mathrm{dyn}}}{\omega}\right)^2, q = \frac{2F_{\mathrm{d}}\left(\frac{\mathrm{d}^2 s}{\mathrm{d}z^2}\right)_{\mathrm{dyn}}}{\omega^2 J_{\mathrm{dyn}}},$$

$$Q = \frac{4d}{\omega^2} \quad \text{and} \quad Q_1 = \frac{4F_{\mathrm{d}}}{\omega^2 J_{\mathrm{dyn}}}\left(\frac{\mathrm{d}s}{\mathrm{d}z}\right)_{\mathrm{dyn}}.$$

Equation (6.2.21) reduces to

$$\frac{\mathrm{d}^2\psi}{\mathrm{d}z^2} + (a - 2q\cos 2z)\psi = Q_0 + Q_1\cos 2z \tag{6.2.22}$$

which is the standard form of inhomogeneous Mathieu's equation. After solving the constant term in the steady-state solution of ψ, the dynamic equilibrium position is found out. As $\mathrm{e}^{2iz} = \cos 2z + i\sin 2z$, the real part of the steady-state solution of the complex equation

$$\frac{\mathrm{d}^2\psi}{\mathrm{d}z^2} + (a - 2q\cos 2z)\,\psi = Q_0 + Q_1\mathrm{e}^{2iz} \tag{6.2.23}$$

gives the actual steady-state solution of (6.2.22).

The steady-state solution of (6.2.23) can be found out from standard texts and can be expressed as follows:

$$\begin{aligned}\psi_s = \frac{Q_0}{\Delta}&\left[\sum_{l=-\infty}^{\infty} C_{2l}\mathrm{e}^{-2ilz}\sum_{l=-\infty}^{\infty}\frac{C_{2l}}{2il+\mu}\mathrm{e}^{2ilz}\right.\\&\left.+\sum_{l=-\infty}^{\infty} C_{2l}\mathrm{e}^{2ilz}\sum_{l=-\infty}^{\infty}\frac{C_{2l}}{2il+\mu}\mathrm{e}^{-2ilz}\right]\\&+\frac{Q_1}{\Delta}\left[\sum_{l=-\infty}^{\infty} C_{2l}\mathrm{e}^{-2ilz}\sum_{l=-\infty}^{\infty}\frac{C_{2l}}{2i(l+1)+\mu}\mathrm{e}^{2i(l+1)z}\right.\\&\left.+\sum_{l=-\infty}^{\infty} C_{2l}\mathrm{e}^{2ilz}\sum_{l=-\infty}^{\infty}\frac{C_{2l}}{2i(l-1)+\mu}\mathrm{e}^{-2i(l-1)z}\right]\end{aligned} \tag{6.2.24}$$

where μ and Δ are two constant quantities. The coefficients C_{2l} are to be determined from the solution of the corresponding homogeneous Mathieu's equation's solution. When a and q are known, μ can be determined from the following relation:

$$\cos(i\pi\mu) = \cos(\pi\sqrt{a}) + \frac{\pi q^2 \sin(\pi\sqrt{a})}{4\sqrt{a}(a-1)} \tag{6.2.25}$$

From the homogeneous Mathieu's equation, the following recurrence relation relating the coefficients C_{2l} is obtained:

$$\left[a-(2l-i\mu)^2\right]C_{2l} \quad q\left[C_{2(l-1)}+C_{2(l+1)}\right]=0 \tag{6.2.26}$$

Equating the determinant of the infinite number of coefficients to zero leads to (6.2.25). C_{2l} decreases very rapidly as l in cases only a limited number of terms need to be considered, and solving the simultaneous equations, the coefficients can be determined. The constant part of the solution can be expressed as

$$\psi_{\text{sconst}} = \frac{Q_0}{\Delta}H_0^{Q_0} + \frac{Q_1}{\Delta}H_0^{Q_1}. \tag{6.2.27}$$

But $\underset{t\to\infty}{\delta t}\ \frac{1}{t}\int_0^t \psi \mathrm{d}t = 0$ the steady-state solution of ψ cannot have a nonzero value, and therefore, $\psi_{s_{\text{const}}} = 0$ or

$$Q_0 H_0^{Q_0} + Q_1 H_0^{Q_1} = 0. \tag{6.2.28}$$

The recursive equation yields the quantities $H_0^{Q_0}$ and $H_0^{Q_1}$ in the following form

$$H_0^{Q_0} = 2\left[\frac{C_0^2}{\mu} + C_2C_{-2}\left(\frac{1}{\mu+2i}+\frac{1}{\mu-2i}\right) + C_4C_{-4}\left(\frac{1}{\mu+4i}+\frac{1}{\mu-4i}\right)+\cdots\right]$$

and

$$H_0^{Q_1} = \left(\frac{C_0 C_{-2}}{\mu} + \frac{C_2 C_0}{\mu + 2i} + \frac{C_{-2} C_{-4}}{\mu - 2i} + \frac{C_4 C_2}{\mu + 4i} + \frac{C_{-4} C_{-6}}{\mu - 4i} + \cdots \right)$$
$$+ \left(\frac{C_0 C_2}{\mu} + \frac{C_2 C_4}{\mu + 2i} + \frac{C_{-2} C_0}{\mu - 2i} + \frac{C_4 C_6}{\mu + 4i} + \frac{C_{-4} C_{-2}}{\mu - 4i} + \cdots \right)$$

It is to be remembered that a, q, Q_0 and Q_1, all depends on $\theta_{2_{\text{dyn}}}$ and are still undetermined. If the shift is χ,

$$\chi = \theta_{2_{\text{dyn}}} - \theta_{2_0} \tag{6.2.29}$$

Using (6.2.29) and expanding a, q, Q_0 and Q_1 in Taylor's series, one obtains the following:

$$a\left(\theta_{2_{\text{dyn}}}\right) = a(\theta_{2_0} + \chi) = a(\theta_{2_0}) + \left(\frac{\mathrm{d}a}{\mathrm{d}\theta_2}\right)_{\theta_{2_0}} \chi + \cdots = a_0 + a_0' \chi + \cdots$$

$$q\left(\theta_{2_{\text{dyn}}}\right) = q(\theta_{2_0} + \chi) = q(\theta_{2_0}) + \left(\frac{\mathrm{d}q}{\mathrm{d}\theta_2}\right)_{\theta_{2_0}} \chi + \cdots = q_0 + q_0' \chi + \cdots$$

$$Q_0\left(\theta_{2_{\text{dyn}}}\right) = Q_0(\theta_{2_0} + \chi) = Q_0(\theta_{2_0}) + \left(\frac{\mathrm{d}Q_0}{\mathrm{d}\theta_2}\right)_{\theta_{2_0}} \chi + \cdots = Q_{00} + Q_{00}' \chi + \cdots$$

$$Q_1\left(\theta_{2_{\text{dyn}}}\right) = Q_1(\theta_{2_0} + \chi) = Q_1(\theta_{2_0}) + \left(\frac{\mathrm{d}Q_1}{\mathrm{d}\theta_2}\right)_{\theta_{2_0}} \chi + \cdots = Q_{10} + Q_{10}' \chi + \cdots$$

But when $\theta_2 = \theta_{2_0}$,

$$\lambda = F_0 \left(\frac{\mathrm{d}s}{\mathrm{d}\theta_2}\right)_{\text{dyn}} = \left(\frac{\mathrm{d}U}{\mathrm{d}\theta_2}\right)_{\text{dyn}} = 0$$

and, thus,

$$Q_{00} = \frac{4d}{\omega^2} = \frac{4\lambda}{\omega^2 J_{\text{dyn}}} = 0$$

As it is expected that χ is a small quantity, one can use the approximations

$$a \approx a_0, q \approx q_0, Q_{00} \approx Q_{00}' \chi, Q_1 \approx Q_{10}$$

without introducing too much error. Hence, the equation of motion with these approximations can be written as follows:

$$\frac{d^2\psi}{dz^2} + (a_0 - 2q_0 \cos 2z)\psi = Q'_{00}\chi + Q_{10}\cos 2z + \cdots \quad (6.2.30)$$

in which a_0 and q_0 are known. So solving the homogeneous equation with these values of a_0 and q_0, one obtains $H_0^{Q_0}$ and $H_0^{Q_1}$ Now,

$$Q'_{00} = \left(\frac{dQ_0}{d\theta_2}\right)_{\theta 2_0} = \frac{4}{J\omega^2}\left(\frac{d\lambda}{d\theta_2}\right)_{\theta 2_0}$$

But again $(d\lambda/d\theta_2)\theta_{2_0} = -K$ and using this in the above equation,

$$Q'_{00} = \frac{-4K}{J\omega^2}$$

Now, using the approximations $Q_{0_0} \approx Q'_{00}\chi$, $Q_1 \approx Q_{10}$, (6.2.28) takes the following form:

$$Q'_{00}\chi H_0^{Q_0} + Q_{10}H_0^{Q_1} = 0$$

or

$$-\frac{4K\chi}{\omega^2 J}H_0^{Q_0} + Q_{10}H_0^{Q_1} = 0$$

Finally, the shift χ can be expressed as follows:

$$\chi = \frac{\omega^2 H_0^{Q_1}}{4H_0^{Q_0}}JQ_{10} \quad (6.2.31)$$

To find out the spread, the inhomogeneous Mathieu's equation is solved using the approximations used in the above discussion. Equation (6.2.22) can be written as

$$\frac{d^2\psi}{dz^2} + (a - 2q\cos 2z)\psi = Q'_{00}\chi + Q_{10}\cdot\frac{1}{2}\left(e^{2iz} + e^{-2iz}\right) \quad (6.2.32)$$

The steady-state solution of the above equation is given by

$$\psi_s = \frac{Q_{10}}{4\sum\limits_{l=-\infty}^{\infty} C_{2l}\sum\limits_{l=-\infty}^{\infty} C_{2l}(2l+\beta)}\sum_{l=1}^{\infty}\left[\frac{2H_0^{Q_{10}}}{H_0^{Q_0}}K\sum_{l=-\infty}^{\infty} C_{2l}\left\{\frac{C_{2(l+n)} + C_{2(l-n)}}{-2l+\beta}\right\}\right.$$
$$\left. - \sum_{l=-\infty}^{\infty} C_{2l}\left\{\frac{C_{2(l-1-n)} + C_{2(l-1+n)}}{-2(l-1)+\beta} + \frac{C_{2(l+1+n)} - C_{2(l+1-n)}}{-2(l+1)+\beta}\right\}\right]\cos 2nz$$
(6.2.33)

where $\mu = i\beta$. If the amplitude of this steady-state oscillation be Λ, then the spread is given by 2Λ.

However, the equation of motion can be solved directly using computers. The effects of energy dissipation mechanisms have not been taken into account, but an approximate solution can be obtained when the resultant effect of the different dissipation mechanism is clubbed together and represented with the help of an equivalent viscous damper. The equivalence is maintained through equalizing the amount of energy dissipation per cycle.

6.3 Dynamics of Mechanisms with Flexible Links

Sometimes, the micromechanism's mobility is obtained through the compliance of the members and not through localized flexibility. Such a mechanism may be monolithic or composed of flexible links connected through kinematic pairs. Considerable study has been conducted on mechanisms that are composed of elastic links through both elastodynamic and kinetoelastodynamic approaches. In case of elastodynamic analysis, the effects of elastic deformation on the change in kinematic dimensions of a linkage, resulting in change in its kinematic parameters such as displacement, velocity and acceleration, are not taken into account. Whereas in kinetoelastodynamic analysis, these effects are used to determine the altered kinematics of a system that in effect again change the subsequent elastic deformations due to altered inertial loading on the links. But in general, the masses of the elements of a micromechanism are very small as mass scales as length dimension^3. Therefore, kinetoelastodynamic analysis is not essential for determining the response of micromechanisms to dynamic loading as inertial loading plays a major role in 'kinetoelastodynamic analysis'. However, kinetostatic and elastodynamic analysis may be useful in certain circumstances.

Although analytical treatment, shown in Sects. 6.2.1 and 6.2.2, is possible in case of micromechanisms with flexure hinges, such a generalized approach is not usually feasible for analysing mechanisms with distributed elasticity (i.e. the links are feasible). Most often, finite element approach is adopted for elastrodynamic and kinetostatic analyses and individual programming needs to be developed for each individual case. Only in special situations where the number of flexible links is very limited and such link(s) possess simple geometry, analytical treatments are possible. In what follows elastodynamic analysis of a simple slider–crank mechanism is presented in which only the connecting rod is a flexible straight link as shown in Fig. 6.5.

In this simplified analysis, the slider is assumed to be subjected to no external force along the slider axis and is considered to possess negligible mass. The crank OA of radius r rotates at a constant angular speed ω. The chosen coordinate system xy is attached to the coupler (connecting rod) AB of length l with its origin at B as shown in the figure. The frequency of natural oscillation of the coupler along the longitudinal axis is very high compared to that in the lateral direction. Thus, only the elastic vibration in the y-direction is of significance.

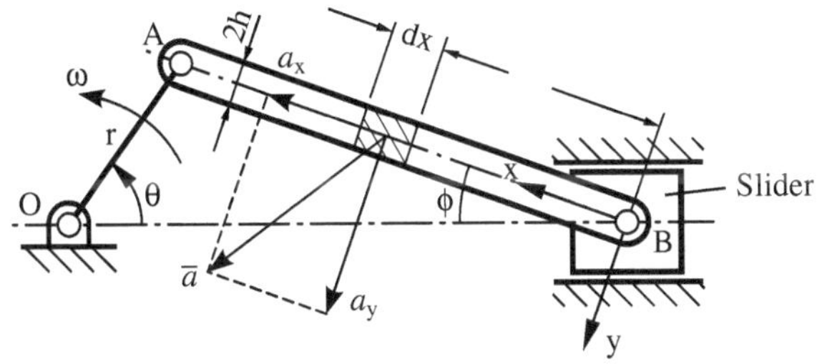

Fig. 6.5 Slider–crank mechanism with a flexible connecting rod

The equation of motion for transverse elastic oscillation of the coupler can be written as follows:

$$EI\frac{\partial^4 w(x,t)}{\partial x^4} + \rho A\frac{\partial^2 w(x,t)}{\partial t^2} = -\rho A a_y. \tag{6.3.1}$$

where *EI* is the flexural rigidity of the coupler, *A* is the cross-sectional area of the coupler that is constant, ρ is the density of the coupler material, $w(x, t)$ is elastic deflection of a point on the coupler axis and a_y is the *y*-component of the acceleration of an element at a distance *x* from the origin at point B (Fig. 6.5). Using the non-dimensional parameters $E = \omega t$, $\bar{x}=x/l$, $\bar{w} = w/l$ and $q = a_y/\omega^2 l$, the equation can be written as

$$k^4\frac{\partial^4 \bar{w}}{\partial \bar{x}^4} + \frac{\partial^2 \bar{w}}{\partial E^2} = q \tag{6.3.2}$$

where $k^4 = EI/(\rho A\omega^2 l^4)$. Again considering the kinematics of the mechanism, a_y can be written in the following form:

$$a_y = \omega^2 r\sin(\theta+\varphi) + (l-x)\frac{d^2\phi}{dt^2}$$

Expending the R.H.S of the above equation in series form, one gets

$$q = -\sum_{n=1}^{\infty}(A_n + B_n\bar{x})\sin n\theta \tag{6.3.3}$$

where A_n and B_n are functions of only r/l ($=\lambda$). Again, $(A_n + B_n\bar{x})$ can be expanded in a Fourier series in $\bar{x}$ and the expression for q takes the following form:

$$q = -\sum_{n=1}^{\infty}\frac{2}{\pi}\left[(2A_n + B_n)\sum_{s=1}^{\infty}\frac{\sin(2s-1)\pi\bar{x}}{2s-1} - B_n\sum_{s=1}^{\infty}\frac{\sin 2\pi s\bar{x}}{2s}\right]\sin n\theta.$$

Using the above expression for the solution of (6.3.2) can be written in the following form:

$$\bar{w}(\bar{x},\bar{t}) = \sum_{n=1}^{\infty}\sum_{s=1}^{\infty}\left[\alpha_{ns}\sin(2s-1)\pi\bar{x} + \beta_{ns}\sin 2s\pi\bar{x}\right]\sin n\bar{t}$$

and the coefficients α_{ns} and β_{ns} can be expressed as follows:

$$\alpha_{ns} = \frac{-\frac{2}{\pi}\left(\frac{2A_n+B_n}{2s-1}\right)}{\pi^4(2s-1)^4k^4 - n^2},\ \beta_{ns} = \frac{\frac{B_n}{\pi s}}{\pi^4(2s)^4k^4 - n^2}$$

After obtaining the solution for $\bar{w}$, the bending stress in the coupler can be estimated from the following relation:

$$\sigma_b = \pm\frac{Eh}{l}\frac{\partial^2\bar{w}}{\partial\bar{x}^2} \tag{6.3.4}$$

where $2h$ is the thickness of the coupler as shown in Fig. 6.5.

It is clear from the above example that elastodynamic analysis of mechanisms is extremely complex when analytical solutions are sought. Computational approach is, therefore, preferred.

Chapter 7
Microactuators

Abstract Microactuators are essential for operating micromechanisms. Design of microactuators becomes successful only when the scaling laws are taken into consideration. The presentation shows how the scaling law can be advantageously employed. Different principles used in developing microactuators are discussed and a generalized comparison of different types of microactuators is presented. Electromagnetic and electrostatic microactuators are discussed and their principle of operation is explained. Use of smart materials for actuation purpose can be very effectively utilized for developing many microactuators. The use of piezoelectric materials, shape memory alloy, electroactive polymers (EAPs) and some magnetostrictive materials is discussed.

Keywords Electromagnetic microactuators · Electrostatic microactuators · Functional material-based microactuators · Electroactive polymers · Actuation mechanisms

7.1 Basic Principles and Considerations

For accomplishing the physical functions, a micromechanism needs to have the ability to make mechanical movements in a controlled fashion. Intrinsically, a micromechanism does not possess this ability and suitable drives (or, actuators) are necessary for a mechanism to execute physical movements. As it has been noticed during the discussions on design and conceptualization of micromechanisms, in case of microactuators, also many special considerations are essential for successful operation. Though a very large amount of R&D has been carried out in the area of microactuators, a systematic study of the subject in a comprehensive manner is still lacking. All possible types of phenomena have been considered for their gainful implementation in developing microactuators and bewildering variety of devices emerged from these efforts. However, only some have resulted in commercial products and a majority is still within the walls of research laboratories. This chapter is devoted to a discussion of the important developments.

A. Ghosh and B. Corves, *Introduction to Micromechanisms and Microactuators*, Mechanisms and Machine Science 28, DOI 10.1007/978-81-322-2144-9_7

7.1.1 Scaling Effects

A direct scaling down of a macrosized actuation device generally does not lead to a successful microactuation device because of the scaling laws. In Chap. 3, the matter of scaling laws has been discussed in details and it has been shown how different physical phenomena can take the predominant roles at different sizes. For successful design of microactuators, it is essential to take the scaling laws into consideration. Sometimes, the scaling effects can be gainfully utilized in designing microactuators.

As surface-related forces scale as l^2 and those related to volume scale as l^3 at microscale, surface-related forces dominate. This is the reason why microtribological effects and phenomena are so important to be considered while designing microsystems with bodies in relative motion. Table 3.1 shows the scaling of different types of forces. It is clear from the table that elastrostatic force predominates at microscale as it scales as l^2, whereas electromagnetic forces scale as l^4 (or, l^3). Hence, it is natural that for microactuators, electrostatic phenomena are more suitable.

It is also not difficult to show how the scaling effect can be effectively used in enhancing the performance of a microactuator. An example can demonstrate the matter clearly. Figure 7.1a shows an electrostatic actuator with a characteristic length L, and the actuator develops a force F.

If the actuator is downscaled so that the characteristic length of the microactuator is L/n, the force this miniaturized version will develop will scale down as n as electrostatic force scales as l^2 (Fig. 7.1b). Thus, such a microactuator can develop a force $F/n^2(=\mathrm{F} \cdot \{(\mathrm{L/n})/\mathrm{L}\}^2)$. The total volume of the original actuator is proportional to L^3 (as volume scales as l^3) and that of the microactuator is $(\mathrm{L/n})^3$. Thus, it is possible to pack approximately n^3 microactuators in the volume occupied by the original actuators. If all these microactuators are linked to work in parallel, the total developed force becomes

$$\left(\frac{F}{n^2}\right) \times n^3 = nF$$

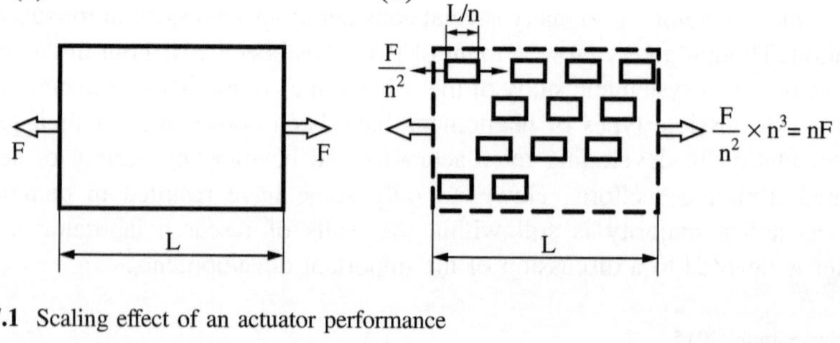

Fig. 7.1 Scaling effect of an actuator performance

Thus, when n is large, the above massively parallel system with miniaturized actuators can develop a much larger force than that developed by a macrosized actuator. In fact, this phenomenon is used by nature for developing the actuators (i.e. muscles). The muscles are composed of very large number of miniaturized actuation devices all connected to function in parallel. This is the way to develop the capability to develop large forces.

Scaling effects on surface-related forces tend to dominate the scene when microparts slide against one another. Adhesion and stiction are important problems that need special attention. Unfortunately, this problem still remains as a serious hurdle in developing some types of microactuators. Since it has not been possible to develop ball bearings and other types of anti-friction devices at that scale and using fluid films create their own problems, the only option available is to use flexure hinges for the necessary support. But that limits the range of movement and may not be acceptable as alternative solutions.

7.1.2 Design Considerations

Microactuators have been designed and conceived using a large variety of physical phenomena and, as a result, there exists a bewildering variety of these devices. It is extremely difficult to develop a systematic classification of all types of microactuators; however, such a classification can help in the study of these devices in the field of emerging technology of microsystems. A majority of the existing microactuators can be classified mainly from two points of view. One classification is based on the basic characteristic of their physical construction and the other can be based upon the principle on which the actuators function.

In the first type of classification, all microactuators can be divided into two classes. In one group, the primary mechanical motions produced are in the form of rigid body motion of the relevant elements. In the other group, mechanical movements are results of deformations generated in relevant members of the device. Figure 7.2 shows this primary classification.

The microactuators in the first group are composed of links connected through kinematic pairs such as conventional macroscopic machines. Because of the very nature of these systems, relative movements exist between neighbouring elements. This makes microtribology a very important aspect of such types of microactuators, and adhesion and stiction can be serious problems in these cases. Mounting and supporting the freely moving members in these microactuators is also a problem as developing anti-friction bearings, and mountings at that scale is still technologically not feasible. To overcome the problem, in many cases, the moving members are supported by flexure hinges. Figure 7.3 represents such approach. However, this imposes limits on the range of movement of the component under consideration.

In the second group, the mechanical movement of the desired member is not generated by some motion of a link or member, driven by some force/moment

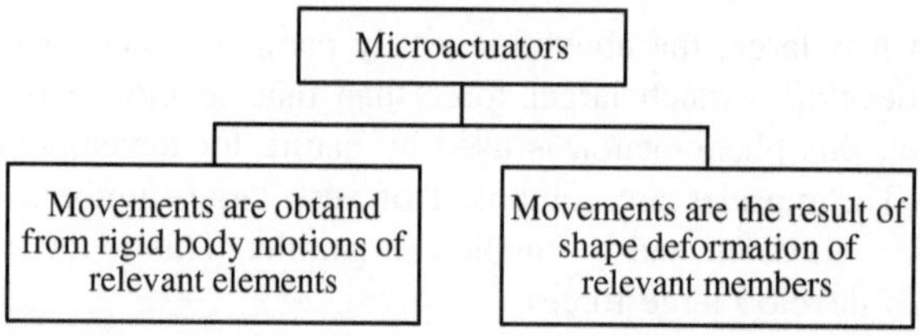

Fig. 7.2 Scaling effect of an actuator performance

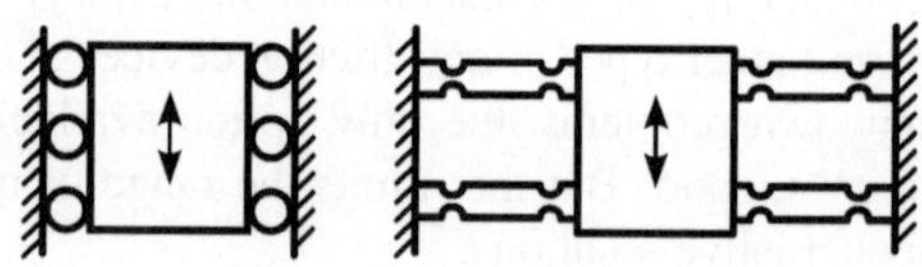

Fig. 7.3 Replacement of roller guides by flexure hinges in micromechanisms

generated due to applied input electrical signal, but the movement at the required location is a result of the shape deformation of some members when the input signal is present. Figure 7.4 shows schematically a few such possibilities. There are a number of physical phenomena that can cause such deformations in a solid element, the most common ones being piezoelectric effect, magnetostriction, shape memory effect, etc.

More recently, ionic metal-based polymer composites have been developed which can produce quite large deformation when a relatively small voltage is applied. In fact, such smart materials are used for developing artificial muscle like objects for manipulation and locomotion. It is needless to emphasize that in case of these groups of actuators, the range of movement is limited. On the other hand, such microactuators are generally less affected by microtribological phenomena and are also easier to integrate with the whole micromachine. In some types of applications such as gripping microscopic objects, deformation microgrippers can be easy to design and fabricate. Such deformations are produced either by using combination of materials with different characteristics produce deformation or by using functional (or, smart) materials which respond to external stimulation by changing shape. Figure 7.5 shows these.

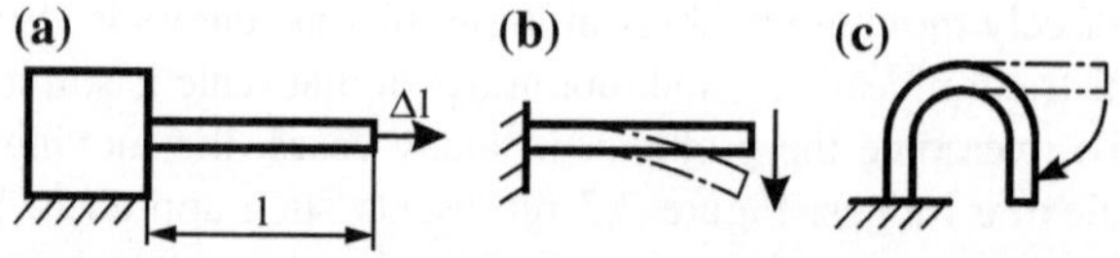

Fig. 7.4 Actuation by shape deformation. **a** Electromagnetic field produced by a conductors element, **b** electromagnetic field produced by a permanent magnet, **c** deformation of element due to shape memory effect

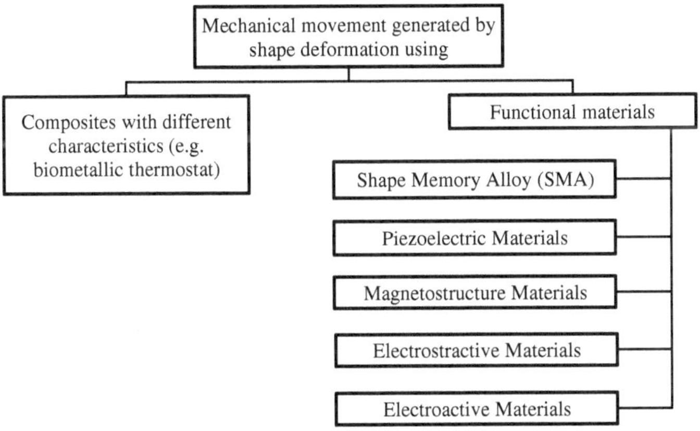

Fig. 7.5 Classification of actuation based upon shape deformation

The most common types of microactuators are shown in Fig. 7.6.

Correct choice of microactuator for a particular application needs careful consideration of various characteristics of the available microactuation principles. Table 7.1 represents the common types of actuation mechanisms and some important features for easy comparison.

Important parameters and important characteristics are presented in Table 7.2 to facilitate easy comparison.

There are some other types of very specialized microactuators, but the table covers the most commonly used type actuators. While selecting a suitable microactuator for a specific application, one also needs to keep in mind the fabrication of the device and some special requirements like biocompatibility. Material and working voltage play very important roles in this matter.

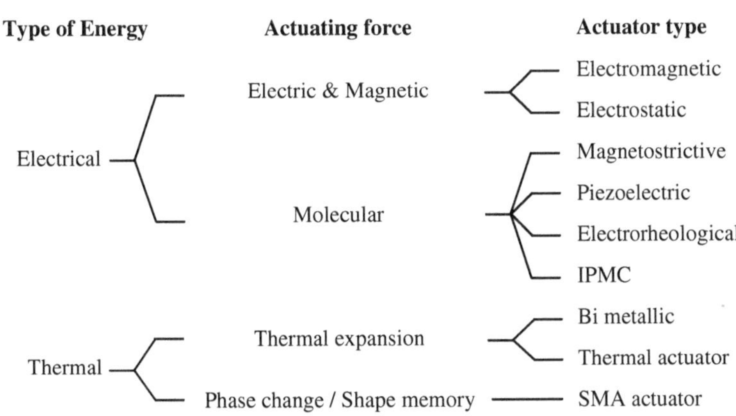

Fig. 7.6 Types of microactuators

Table 7.1 Common types of actuation mechanisms

Actuation principle	Transformation	Important feature
Electromagnetic and electrostatic	Electricity/ magnetism → mechanical force	Motor size varies between 10 and 100 μm
Piezoelectric	Electricity → mechanical strain Mechanical strain/force	Quick response, high force, small displacement
SMA	Heat → mechanical strain/force	High deformation and slow response
Electrorheological	Electricity → change in viscosity Viscosity	High output, valve operation, flow control
Thermal	Heat → strain/deformation	Slow response, low force/output
Electrochemical	Electricity → mechanical deformation	Complex shapes, fluid environment, large displacement
Photomechanical	Light → force	Polymer based

7.2 Electric Microactuators

This section is going to be devoted to a brief discussion on different commonly used microactuators. Working principles and the mechanics of the processes involved will be presented keeping the level introductory in nature. Special points relating to the fabrication technique, if any, will be discussed in the chapter on microfabrication.

7.2.1 Electromagnetic Microactuators

In the macroscopic scale, driving systems based on electromagnetism are most commonly used. But due to the unfavourable scaling laws, most of these devices become ineffective when miniaturized. Even then, there have been attempts to develop microsized motors and actuators working on the principle of electromagnetic induction. The efficiency of conversion of electrical energy into mechanical energy is high in case of electromagnetic devices. The operation of electromagnetic devices is also quite smooth and response is very fast! But as mentioned earlier, the downward scalability of electromagnetic actuators into the micro- and nanorealms is poor. Apart from this, there are some technological problems in fabricating microsized coils. More importantly, most electromagnetic actuators require perpendicularity between the current-carrying conductors and the moving element that is difficult to realize through standard planar micromachining techniques.

In spite of the difficulties electromagnetic rotary microactuators and linear actuators have been developed. Figure 7.7 shows a motor that can develop up to 7.5 μN at a speed of 100,000 rpm. The diametrically magnetized rare earth permanent magnet forms the rotor.

Table 7.2 Comparison of different types of microactuators

Primary source of input	Actuators	Kinematic character	Range of motion	Developed force/torque	Response	Operating voltage	Power density
Electromagnetic induction	Motors	Rotation	Large	Small	Fast	Low	High
Static electricity	Rotary motors	Rotation	Large	Small	Fast	High	Low
	Linear actuators	Rectilinear motion	Small	Medium	Fast	High	Low
Piezoelectricity	Laminated actuators	Rectilinear motion	Small	Large	Fast	High	High
Heat/thermal expansion/phase change	SMA actuators	Linear/angular motion	Large	Medium	Medium	Low	Very high
	Thermal expansion	Linear	Small	Large	Slow/medium	Low	Medium
	Bimetallic strips	Angular	Medium	Medium	Medium	Low	Medium
Electroactive polymeric effect	IPMC strips	Angular	Large	Small	Fast	Low	Low

The soft magnetic outer cylindrical tube acts as housing and also as the return yoke for the magnetic flux. Apart from this longitudinal motor, a button-shaped 1.4-mm-thin penny motor has been developed that has a diameter of 12.8 mm. It can develop 80 μNm at a speed of 60,000 rpm and draws 200 m current. Another permanent magnet microactuator that has been developed using a scheme is shown in Fig. 7.8.

The rotor is a small cylindrical permanent magnet of 1 mm diameter and 0.5 mm thickness. This rotor is made of SmCo 4-pole radially magnetized permanent magnet. The coil is 1 mm in diameter and 2.5 mm in length and has about 1,000 turns. Using miniaturized electromagnetic vibrators, innovative drives have been developed by converting the vibration into linear motion. However, as the motion is sensitively dependent upon the surface characteristics of the plane on which it moves, it is difficult to accurately control the motion.

There are two basic principles followed for developing electromagnetic microactuators. Figure 7.9 shows these schematically. In the first case, shown in Fig. 7.9a, the electromagnetic force is developed between two conductors carrying currents I_1 and I_2.

This is due to the magnetic field B generated by one conductor (say 1) that is kept fixed. The other conductor 2 is mobile and produces the mechanical motion necessary in an actuator. In the second case, the magnetic field B is produced by a permanent magnet as shown in Fig. 7.9b. In the first case,

$$B = \frac{\mu_0}{2\pi r} I_1 \tag{7.2.1}$$

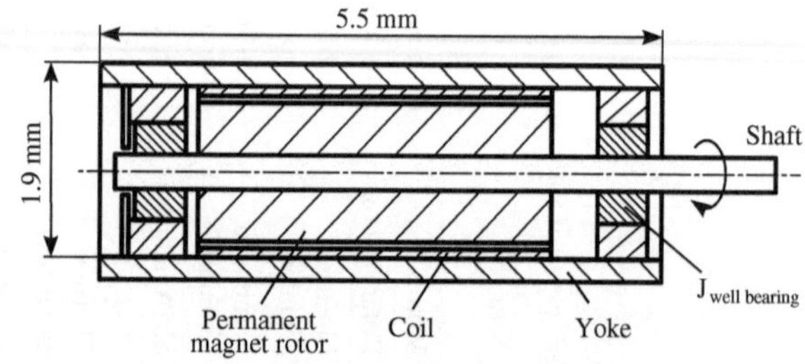

Fig. 7.7 Electromagnetic linear microactuator

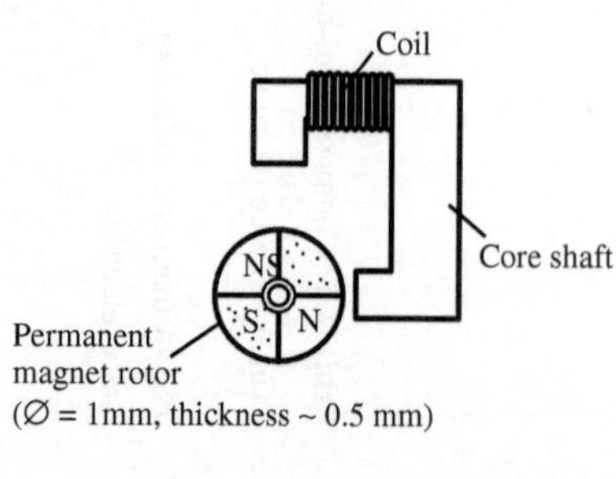

Fig. 7.8 Permanent magnet rotary microactuator

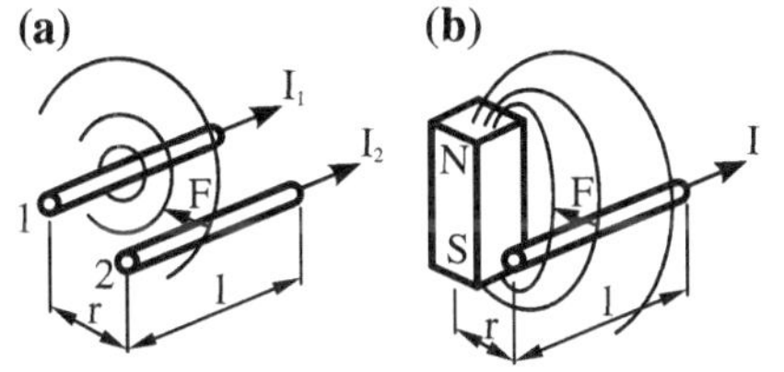

Fig. 7.9 Principles of electromagnetic microactuation. **a** Elecromagnetic field produced by a conductors element, **b** electromagnetic field produced by a permanent magnet

where μ_0 is the permeability and r is the distance for the conductor 1. If the length of the conductor 2 be l, the force acting on this conductor is

$$F = BlI_2 = \frac{\mu_0 l}{2\pi r} I_1 I_2 \tag{7.2.2}$$

In case of the second configuration, the force on the mobile conductor can be expressed as

$$F = BlI_2 \tag{7.2.3}$$

where B is the magnetic field strength at a distance r from the magnet as shown. To see how these quantities scale with size three different situations may be considered. In the first situation, the maximum current density is kept constant at the maximum permitted value, say $J_{\max}$. Then, the maximum current through a conductor of cross-sectional area A is given by $I = J_{\max} A$. This clearly shows that I scales as l^2. In the second situation, the heat dissipation rate Q/A_s and the specific resistivity of the conductor material be specified than

$$\dot{Q} = \rho \frac{l}{A} \cdot I^2 \tag{7.2.4a}$$

or

$$\frac{\dot{Q}}{A_s} = \rho \frac{l}{A \cdot A_s} \cdot I^2 \tag{7.2.4b}$$

where A_s is the total area of the conductor. The above equation shows that in this case I scales as $l^{1.5}$. Finally, in the third situation, the maximum temperature rise ΔT may be kept to a specified value. Then for a given conductor material for thermal equilibrium,

$$\dot{Q} \propto kA_s \frac{\mathrm{d}T}{\mathrm{d}z} \tag{7.2.5}$$

where k is the thermal conductivity and z represents a direction normal to A_s and along the direction of heat flow. Again, it is known from (7.2.4a)

$$\dot{Q} = \rho \frac{l}{A} \cdot I^2$$

Combining this with (7.2.5)

$$\rho \frac{l}{A} \cdot I^2 = kA_s \frac{\Delta T}{\Delta z} \tag{7.2.6}$$

Therefore, in this situation, I scales as l^1. If a permanent magnet is used instead of a current-carrying conductor to produce the magnetic field, then the force scaling will be different from the first case with a current-carrying conductor. In case of a permanent magnet, the magnetic field B scales as l^0 when both the volume of the magnet and the location are scaled. In the first case, it is clear that B scales as (I/r), that is l^1 or $l^{0.5}$ or l^0 depending on the condition imposed. The scaling of force is shown in Table 7.3 for both types of actuators for three different conditions imposed.

It is clear from the above table that electromagnetic micromotors based upon permanent magnets have a favourable scaling effect as compared to those using electromagnets. Thus, high-strength permanent magnet-type motors have been developed with diameters as small as 2 mm developing stall torque of 10 μNm at 10^5 rpm. Furthermore, permanent magnet-based actuators are simpler in design also. Another important criterion, the achievable power density, also favours the use of permanent magnets in place of coils. With permanent magnets of high strength, much higher flux density can be achieved as the system becomes independent of thermal limitations.

In view of the fact that continuous unlimited rotational movement is possible, rotary micromotors are favoured in many applications. Because of the convenience with which lithographic microfabrication techniques can be used disc-type axial

Table 7.3 Scaling of force in electromagnetic actuators

Condition	Scaling of I	Scaling of F		
		Current–current configuration $F = \frac{\mu_0 l}{2\pi r} I_1 I_2$	$F = BlI_2$ Scaling of B	Current–magnet configuration
Maximum allowable current density	l^2	l^4	l^0	l^3
Specified heat dissipation rate	$l^{1.5}$	l^3	l^0	$l^{2.5}$
Specified temperature rise	l^1	l^2	l^0	l^2

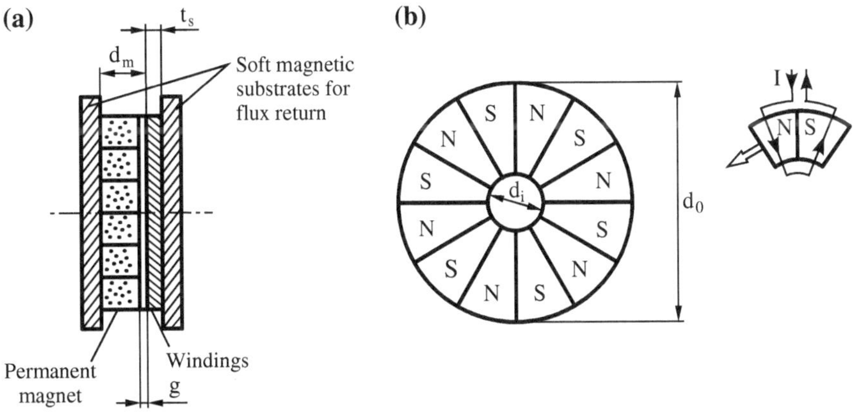

Fig. 7.10 Simple micromotor. **a** Axial flux permanent magnet micromotor, **b** stator winding configuration near a pole pair

flux, permanent magnet-type microactuators are preferred. Figure 7.10a shows a simple motor of this type.

Figure 7.10b shows one stator winding adjacent to a pole pair schematically. When the multiphase stator windings are supplied with a multiphase current, a rotating magnetic field is generated. The rotor permanent magnets are pulled to align with this rotating stator field causing the permanent magnetic disc to rotate accordingly. The dimensions of a typical axial flux permanent magnet micromotor are as follows:

Outer diameter (d_o)	1,000 μm ≡ (mm)
Inner diameter (d_i)	400 μm
Air gap (g)	30 μm
Stator layer thickness (t_s)	160 μm
Thickness of permanent (t_m)	200 μm
Pole pairs	6
Turns in stator windings	12

This motor draws 11 mA current and develops a maximum torque of 0.2 μNm. Though, from the beginning of the development of microsystems and microactuators, electrostatic actuators have been preferred, slowly electromagnetic microactuators are gaining acceptability in the community.

7.2.2 Electrostatic Actuators

In the macroscopic scales, the electrostatic forces are relatively small; besides, very high operating voltages are required. Therefore, the use of electrostatic forces for actuation purpose is rare in the macroscopic scale. On the other hand, electrostatic actuation is very attractive for microsized devices; and, from the very beginning, electrostatic actuators have found favour with the technologists.

Electrostatic actuators have been developed for both rectilinear and rotary motions. The basic principle is the coulomb attraction between oppositely charged bodies (or repulsion between bodies with similar charges). Figure 7.11 shows two electrically charged plates with opposite polarities.

If one of the plates is kept fixed, the other will move towards the fixed one as shown. Such parallel plate configuration can be used as an actuator. There are basically two types of parallel plate actuators—(i) gap-closing and (ii) constant-gap actuator. Figure 7.12a, b show the basic principles of these devices.

The principle of operation of gap-closing actuator is quite simple and straightforward. If the two plates carry opposite charges, the free electrode (or actuator) plate 2 will be subjected to an attractive force acting towards 1. The force varies as $1/z^2$ where z is the gap. It is very clear that as the force increases with z decreasing, the situation is unstable and the moving plate will go and hit the stationary plate. To render the device operational with an equilibrium configuration, the movable plate is connected to a spring of stiffness k as indicated in Fig. 7.13.

When no voltage is applied, the force between the plates is zero and the spring attains its natural length when the gap between the two plates is equal to z_0 as shown in the figure. When a voltage V is applied, an electrostatic attractive force is generated and the movable plate moves towards the fixed one. An equilibrium is reached when the attractive force is balanced by the spring force. So if z be the equilibrium position gap corresponding to an applied voltage V

$$k(z_0 - z) = \frac{\lambda V}{z^2} \tag{7.2.7}$$

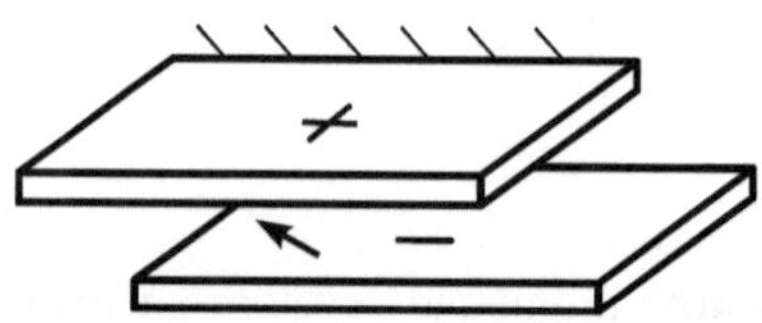

Fig. 7.11 Electrically charged parallel plates in asymmetric position

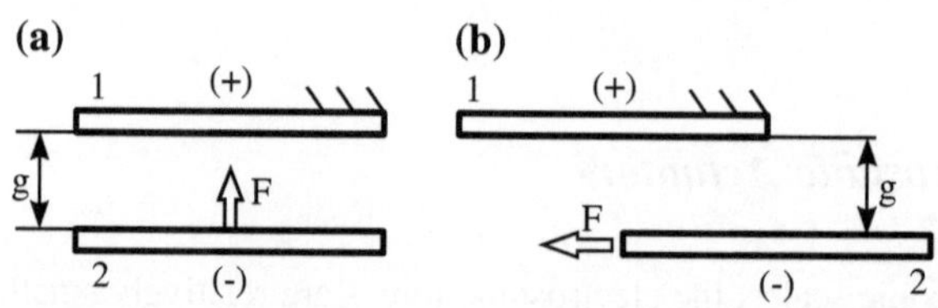

Fig. 7.12 Electrostatic force developed between parallel plates. **a** Symmetric configuration, **b** asymmetric configuration

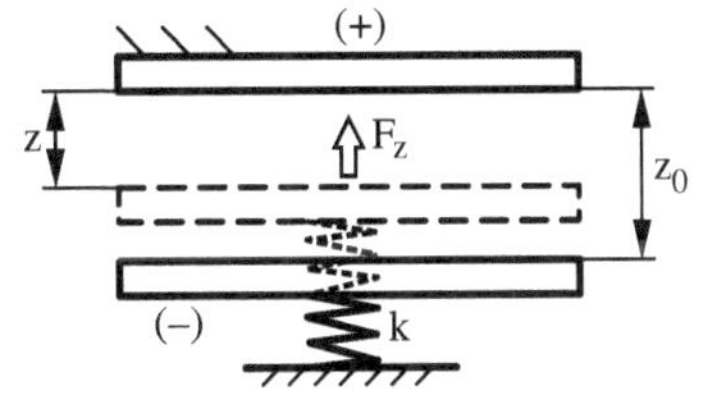

Fig. 7.13 Parallel plate constrained by spring

If the two plates have a surface area A each and subjected to a voltage V the electrostatic energy of the system,

$$U_e = -\frac{\varepsilon_0 A V^2}{2z} \tag{7.2.8}$$

where ε_0 is the permittivity of free space and z is the gap. The strain energy stored in the spring under this condition is

$$U_s = \frac{1}{2}k(z_0 - z)^2. \tag{7.2.9}$$

Thus, the total energy of the system is given by

$$U = \frac{1}{2}k(z_0 - z)^2 - \frac{\varepsilon_0 A V^2}{2z}. \tag{7.2.10}$$

The force involved can be found by using the following relation:

$$F_z = \frac{\partial U}{\partial z} = -k(z_0 - z) + \frac{\varepsilon_0 A V^2}{2z^2} \tag{7.2.11}$$

It is clear that the first term on the RHS represents the restoring force due to the spring and the second term represents the electrostatic attractive force. For an equilibrium to result,

$$k(z_0 - z) = \frac{\varepsilon_0 A V^2}{2z^2} \tag{7.2.12}$$

The second derivative of U can be used to determine the stability of the equilibrium. The equilibrium is stable when

$$\frac{d^2U}{dz^2} \geq 0$$

or,

$$k \geq \frac{\varepsilon_0 A V^2}{z^3}. \tag{7.2.13}$$

When (7.2.13) is used in (7.2.12), the following condition is obtained for the equilibrium position to be stable:

$$\frac{\varepsilon_0 A V^2}{z^3} \leq \frac{\varepsilon_0 A V^2}{2z^2(z_0 - z)}$$

or,

$$z \geq \frac{2}{3} z_0 \tag{7.2.14}$$

This is a very important condition for microactuator design. It implies that only one third of the gap can be used for actuator movement to have stable microactuation (Fig. 7.14).

It is also easy to find out the maximum voltage V_m that can be applied without the movable plate being immediately pulled into the fixed plate (when no fixed steps are provided). Substituting $z = z_0/3$ in (7.2.12) and solving for the voltage yields

$$V_m = \sqrt{\frac{8}{27} \cdot \frac{k z_0^3}{\varepsilon_0 A}} \tag{7.2.15}$$

V_m is also often termed as 'pull-in' voltage.

Using a mechanical stop is not a very elegant solution to prevent 'pull-in' as mechanical stops may be subjected to wear and tear on repeated use, and it is also necessary to maintain the voltage of the stop the same as that of the movable plate. A better solution is to add additional capacitor in series with the microactutators (Fig. 7.15). If the movable plate goes very close to the fixed plate, the capacitance ($=\varepsilon_0 A/z$) will increase.

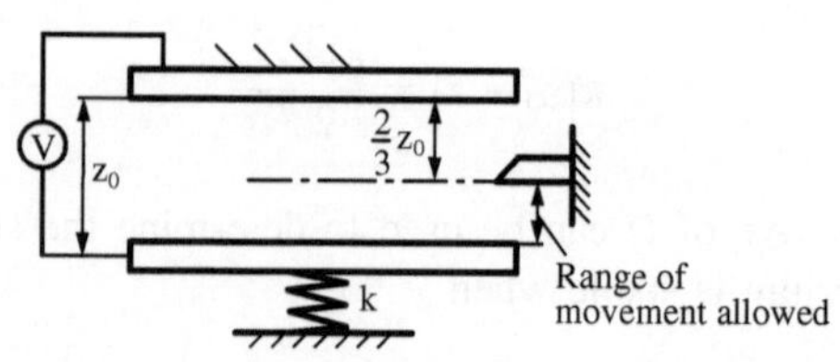

Fig. 7.14 Range of parallel plate movement

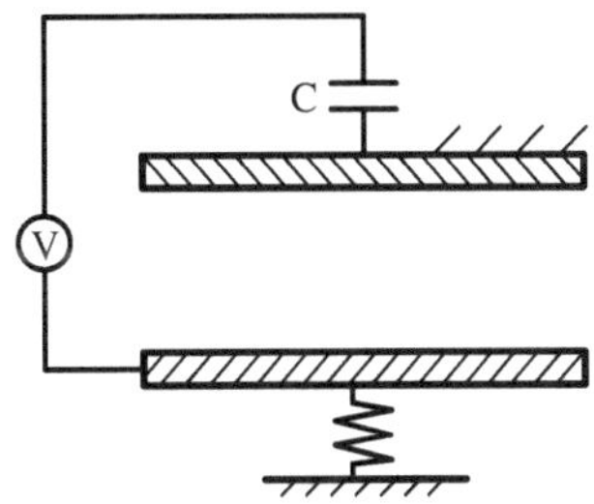

Fig. 7.15 'Pull-in' prevention by a capacitor in parallel plate microactuators

This will cause more voltage to drop against C and the effective voltage on the actuator plates will be less, causing the attractive force to reduce and restoring spring prevents the 'pull-in'. It can be shown that the necessary condition is

$$C \leq 0.5 \frac{\varepsilon_0 A}{z_0}. \tag{7.2.16}$$

With arrangement, the microactuator can be taken to any position in the gap without losing stability. There are many applications of this type of microactuator.

In case of constant-gap plate-type microactuators, the gap between the plates is kept constant. Because of some advantages, the constant-gap electrostatic microactuators are more frequently used. In these devices, the motive force depends on the degree of overlap between the plates (these are also called variable capacitance actuators). Figure 7.16 explains the basic principle of operation and primary characteristics of the plate configuration. When the lower plate is shifted to the right from its aligned position by an amount x, the capacitance of the system is given by

$$C(x) = \frac{\varepsilon_0 t(L - x)}{z_0} \tag{7.2.17}$$

where t is the width of each plate.

When a voltage V is applied, the electric energy U_e is given by the following relation:

$$U_e = \frac{\varepsilon_0 t(L - x)\, V^2}{2 z_0}. \tag{7.2.18}$$

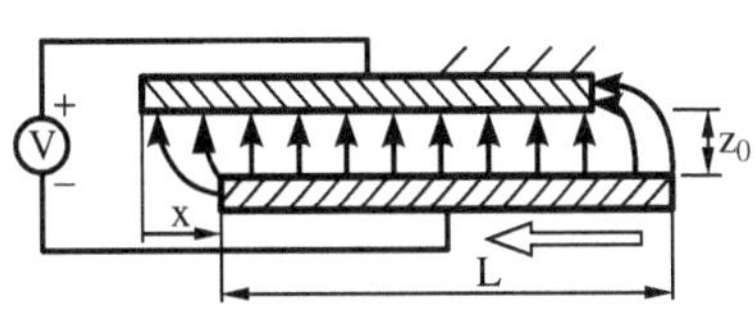

Fig. 7.16 Operational principle of parallel plate microactuator with asymmetric configuration

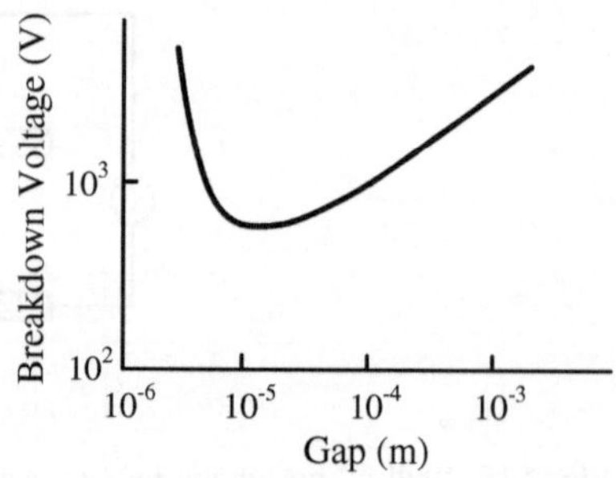

Fig. 7.17 Paschen's curve

The force in the lateral direction is given by

$$F_x = \frac{\partial U_e}{\partial x} = -\frac{\varepsilon_0 t V^2}{2 z_0} \tag{7.2.19}$$

which tries to realign the plates. It is a bit remarkable to note that this force is independent of x but depends only on the gap (constant) and the square of the applied voltage. Keeping z_0 very small, it is possible to increase the force without increasing the voltage too much. The limitation on z_0 is imposed by the breakdown voltage that follows the Paschen's curve shown in Fig. 7.17.

In case of microsystems, fortunately the breakdown voltage shoots up as shown and better performance becomes possible. It is also to be remembered that it is necessary to provide a suitable spring attached to the moving plate.

To increase the driving force, quite often a multiple plate configuration is used as shown in Fig. 7.18. As the device takes the appearance of a comb, these actuators are often called as comb drive actuators.

The force developed in a comb drive (or simply comb) microactuator is

$$F = \frac{n \varepsilon_0 t V^2}{z_0} \tag{7.2.20}$$

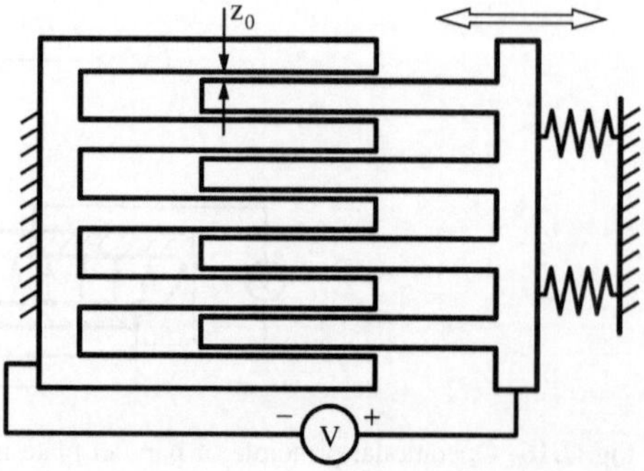

Fig. 7.18 Multiple plate configuration for microactuators (comb actuator)

where n is the number of comb pairs. Typical comb microactuators may have 50 finger with 6 μm width each separated by 1-μm gaps operating at a voltage of 100 V. Such a comb drive can generate up to 13 μN. Comb drives are very popular because of their ease of fabrication. Comb actuators for angular movement have also been developed, and the typical scheme is shown in Fig. 7.19.

Microflexures are used for supporting the moving element that improves the longevity of the devices (both linear and angular) enormously. Comb-type actuators can also act as displacement sensors at the same time because of varying capacitance. This is an additional advantage. Comb drives for transverse motion (gap closing) have also been developed, and the basic scheme is shown in Fig. 7.20. With n elements, the electrostatic force generated is given by

$$F = \frac{1}{2} n \varepsilon_0 z t V^2 \left[\frac{1}{(z_0 - z)^2} - \frac{1}{(z_0 + z)^2} \right] \tag{7.2.21}$$

However, these devices are more popular as sensor due to varying capacitance; its use as actuator is less favoured because of the limited displacement.

Electrostatic micromotors started being developed in the 1980s and opened up new vistas in front of the technologists. The idea of these kinds of microdevices attracted extensive attention from various quarters. The devices are undoubtedly very exotic; however, their use in practice has remained very limited in comparison with the other types of electrostatic microactuators. However, it is expected that

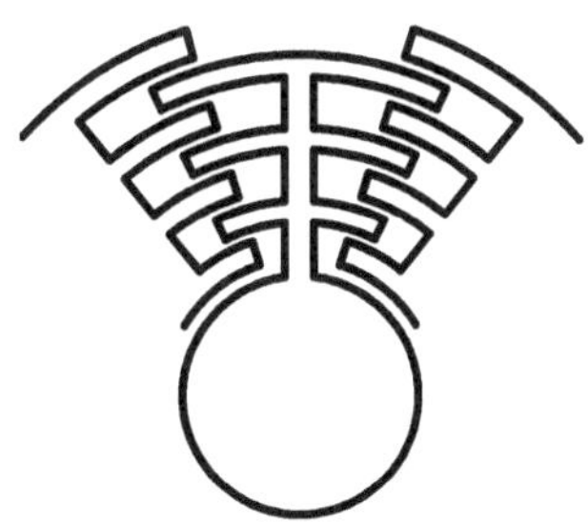

Fig. 7.19 Comb actuator for angular movement

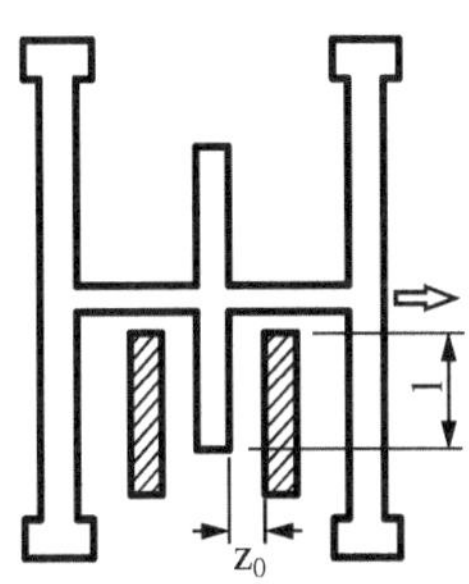

Fig. 7.20 Comb drive for transverse motion

with the advancement of technology, the main tribological issues (due to rubbing motion of the rotor on the fixed hub) can be taken care of.

Electrostatic motors can be classified into mainly three groups as indicated in Fig. 7.21.

The earliest developments resulted in electrostatic motors analogous to stepper motors. The basic construction is shown in Fig. 7.22a. The motor consists of a rotor with several arms which forms a circumferentially varying capacitance with the stator poles.

The rotor is charged and the stator poles are oppositely charged phase wise, one after another. A three-phase electrostatic side-drive motor explains the principle as shown in Fig. 7.22b. The rotor is charged, say, positively and the four start poles in phase 1 are charged negatively. As a result of electrostatic attraction among oppositely charged neighbouring bodies, the rotor rotates to align the rotor to an equilibrium position (when the opposite charges are nearest to each other). Then, phase 2 poles of the stator are charged and the rotor rotates by another step to attain the next equilibrium position. Continuing the process, the rotor can be made to execute stepwise continuous rotation.

The electrostatic torque developed is given by $T = F \cdot R$ where R is the rotor radius. The electrostatic force is given by $\varepsilon_0 w V^2/2g$ where w and g are the width and gap and V is the voltage (Fig. 7.23). Hence,

$$T = \frac{\varepsilon_0 w R}{2g} V^2. \tag{7.2.22}$$

Figure 7.23 shows three types of electrostatic micromotors. Initially, the side-drive electrostatic motors had a disadvantage because of small thickness obtained through lithographic micromachining technique. However, development of LIGA process has made it possible to achieve much higher thickness of micromachined layers. Thus, larger values of w are possible improving the torque generation.

The attempt to develop wobble motors was fuelled by the desire to increase the torque output (and also reduce the microtribological problems). In the axial wobble

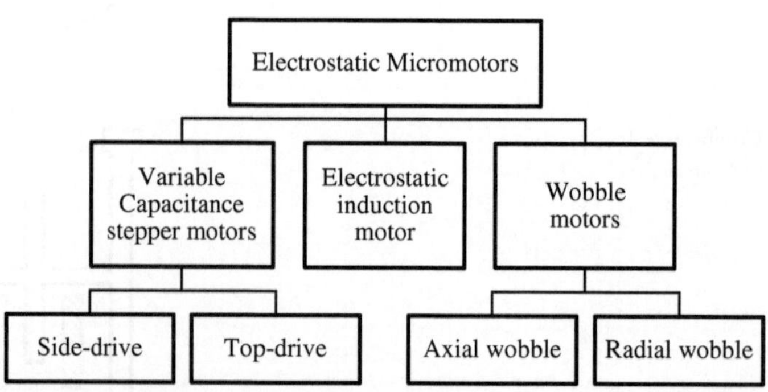

Fig. 7.21 Classification of electrostatic microactuators

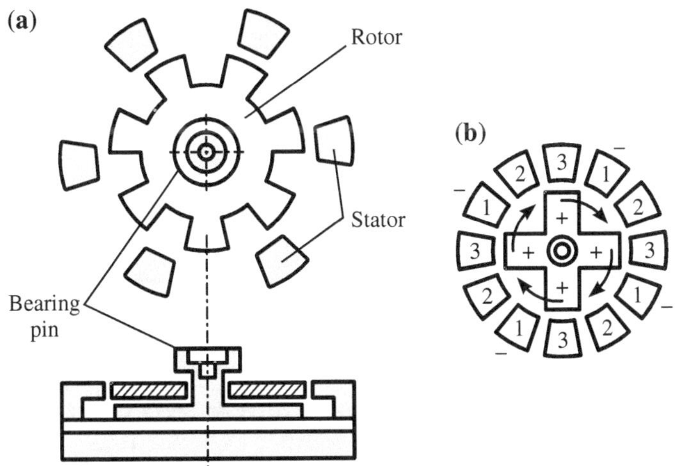

Fig. 7.22 Stepper motor-type electrostatic micromotor. **a** Basic construction, **b** three-phase driving principle

drive motor, the charged rotor disc is attached to a particular stator electrode when it is given an opposite charge. This causes the rotor disc to be inclined and contact with the thin insulating layer covering the charged stator electrode. When the next stator electrode is charged, the rotor disc rolls along and touches the next electrode. Continuing the process, the rotor disc is made to wobble and roll along its circumference. Because of the difference in radius between the rotor and the resulting circle described by the contact point, the rotor is rotated by a small angle after one cycle of activation of all stator poles is complete. If the angular speed with which the stator electrode charge is shifted be ω_s and the angular speed of the rotor be ω_r, the ratio ω_s/ω_r can be termed as a kind of gear ratio (Fig. 7.24).

The radius of the contact point circle can be found out to be

$$R_c = \sqrt{R^2 + d^2} + (h - d)\sin\theta$$

When θ is very small, one can write θ = d/R. With this, one obtains

$$n = \frac{R}{R - R_c} \approx -\frac{2R^2}{2dh - d^2}. \tag{7.2.23}$$

Because of this large effective gearing ratio, large output torques can be obtained. Wear due to sliding friction is also less pronounced because of the rolling nature of the contact. The basic principles are similar in case of radial wobble electrostatic motors. Reduction ratios of up to 200 and output torque of about 200 nNm at 300 rpm have been achieved in case of wobble motors with 560 μm stator diameter.

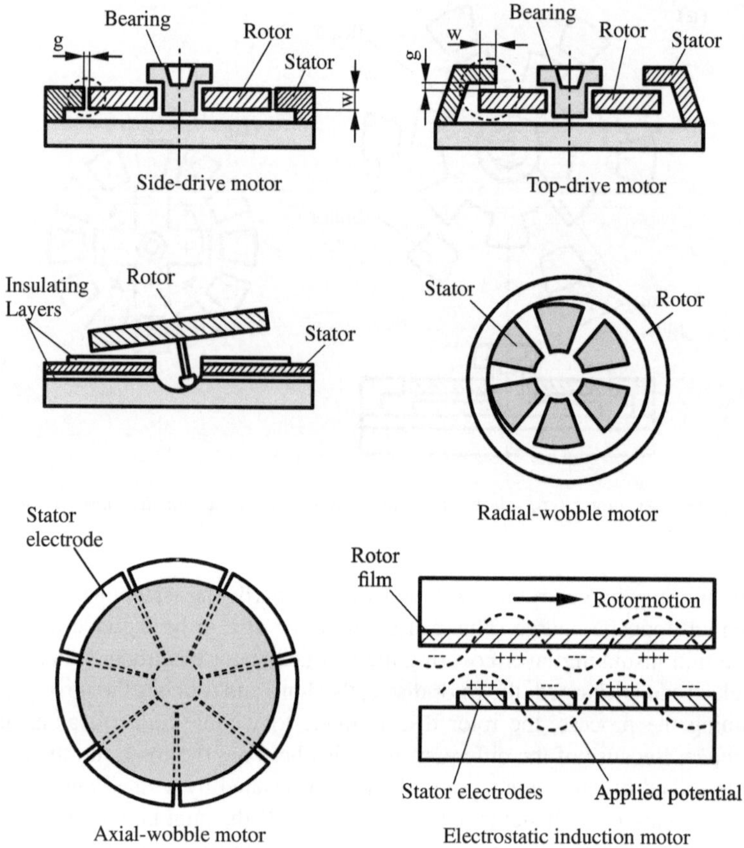

Fig. 7.23 Different types of electrostatic micromotors

Fig. 7.24 Principle of speed reduction

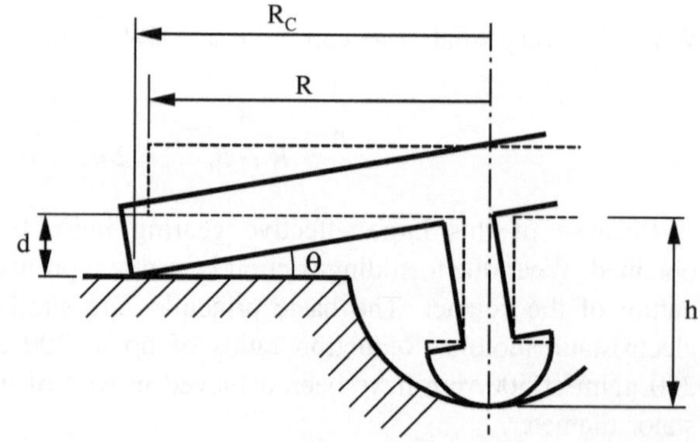

The basic design of electrostatic induction motors consists of two discs separated by a small gap. The rotor disc is coated with a film of conductive material with low conductivity. The stator disc is provided with an array of radial electrodes. By applying a potential wave travelling around the stator, the induced charges on the stator disc become as shown in Fig. 7.23. Electric fields induce charges in the conducting surfaces in the vicinity; fields produced by positive electric charges induce negative charges in the nearby conduct surface and vice versa as shown in Fig. 7.25.

If the conductivity of the rotor is suitably chosen, then the rate at which the induced charge will develop will be slower. Thus, when a travelling electric field is applied, the induced opposite charge will lag as indicated in Fig. 7.23. The tendency of the rotor and charges to align cause the rotor disc to follow the travelling charge distribution in the stator. Since the rate at which the rotor charges develop depends critically on the conductivity of the rotor material, the most important factor in designing induction micromotors is the conductivity of the rotor. In case of variable capacitance motors, the rotor follows the charging of the stator electrodes and the rotation is synchronous with the application of potential to the stator phases. On the other hand, the rotation of the rotor of an electrostatic induction motor is asynchronous. If ω be the electrical frequency and m be the number pole pairs, then the rotor speed Ω is related to the above quantities as follows:

$$S = \frac{\omega - m\Omega}{\omega} \tag{7.2.24}$$

where S is defined as the slip, which is a very important design parameter. When the slip increases initially, the motor torque will increase and then after reaching a maximum, gradually decrease. The rotor resistivity is the deciding factor for giving the value of the critical slip corresponding to the peak torque. The approximate expression for the peak torque is as follows:

$$T_{\text{peak}} \approx \frac{\pi}{4} \frac{m\varepsilon_0 V^2}{1 + \frac{\varepsilon_{SiO_2} g}{\varepsilon_0 d}} \left(R_0^2 - R_i^2\right) \tag{7.2.25}$$

where g is the instantaneous gap, d is the insulation thickness, and R_0 and R_i are the outer and inner rotor radii.

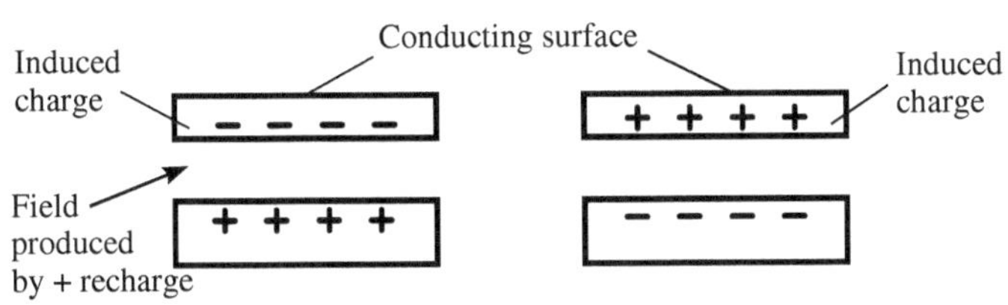

Fig. 7.25 Induced fields on conducting surfaces

7.3 Functional Material-Based Microactuators

Microactuators based on the controlled deformation of elements made up with functional materials (sometimes also called smart materials) generate characteristically small displacements except in the case of electroactive polymers. The main advantage with these types of actuators is that these are free from tribological phenomena. However, these devices are affected by hysteresis and creep. Some important type of such actuators is presented in this section.

7.3.1 Piezoelectric Microactuators

Components made out of piezoelectric materials produce strain when subjected to electric fields. The reverse phenomenon is also observed, that is when the components of piezoelectric materials are subjected to strain, electric fields are developed. The word 'piezo' in Greek means 'to press'. The piezoelectric is produced due to non-centric-symmetric arrangement of atoms in a lattice. Taking crystalline quartz (SiO_2) as an example, the physics is explained through the schematic diagrams shown in Fig. 7.26. It is seen that the arrangements of the positively charged Si atoms and negatively charged O_2 are not symmetric about the centre of the lattice.

When an electric field is imposed, electrostatic forces act on the atoms indicated in Fig. 7.26b, c causing the lattice to stretch or shrink depending on the situation. The strain generated by piezoelectric actuators is typically small (less than 1 %), but developed force can be high. Figure 7.27 shows the schematic diagram of a linear piezoelectric microactuator. Electrodes are integrated onto the upper and lower faces of the piezoelectric film so that the electric field can be applied across the film.

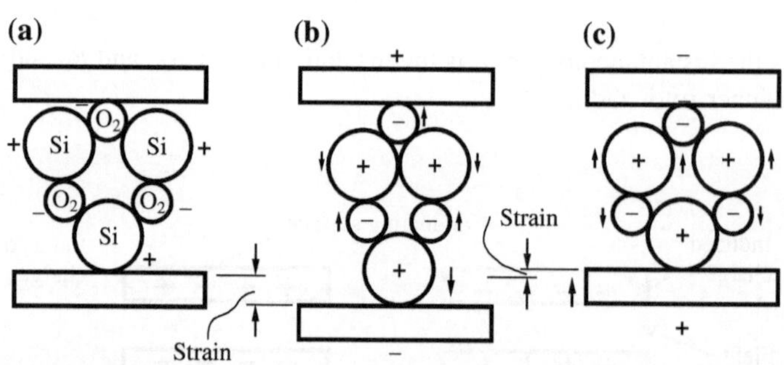

Fig. 7.26 Mechanism of piezoelectric effect in crystalline quartz. **a** With zero applied voltage, **b** strain generated with positive charge at the upper end, **c** strain generated with positive charge at the lower end

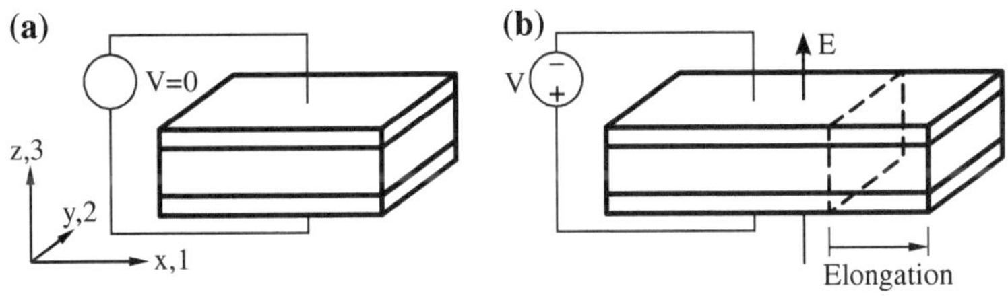

Fig. 7.27 Schematic diagram of a linear piezoelectric microactuator

The electrodes should be thin and compliant so as not to interfere with the dimensional changes in the piezoelectric element. The simplified coupled electrical–mechanical constitutive equations are as follows:

$$S_1 = s_{11}^E \cdot T_1 + d_{31}E_3 \tag{7.3.1a}$$

and

$$D_3 = d_{31} \cdot T_1 + \varepsilon_{33}^T \cdot E_3 \tag{7.3.1b}$$

where S_1 is the x-axis strain, T_1 is the x-axis stress, s_{11}^E is x-axis mechanical compliance at constant electric field, d_{31} is the z-axis to x-axis piezoelectric coupling coefficient, ε_{33}^T is the constant stress z-axis permittivity, D_3 is z-axis electric displacement and E_3 is the z-axis electric field. The important and relevant parameters for some commonly used piezoelectric materials are presented in Table 7.4.

The kind of deformations is explained in Fig. 7.28.

Table 7.4 Relevant piezoelectric properties of some materials

Material	$d_{31} \times 10^{12}$ (C/N)	$d_{33} \times 10^{12}$ (C/N)	$s_{31}^E \times 10^{12}$ (m²/N)	ε_{33}^T
ZnO	−4.7	12	6	8.2
PVDF	−23	−35	333	4
PZT	−171	212	16.4	1,700

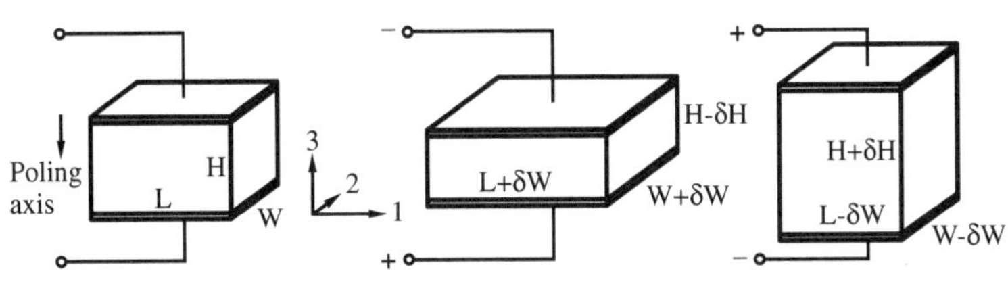

Fig. 7.28 Types of deformation of a piezoelectric crystal

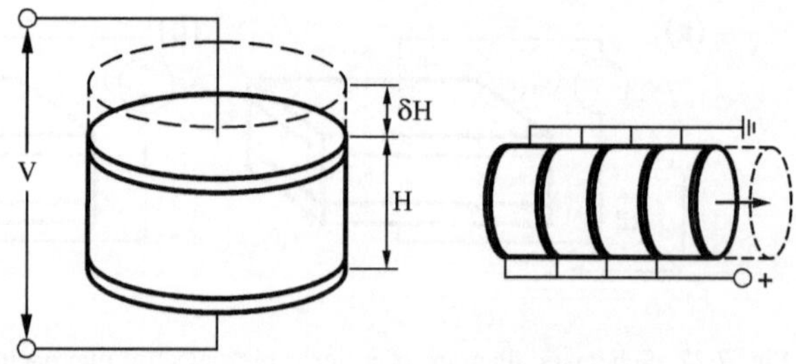

Fig. 7.29 Multiple disc piezoelectric actuator

The various strains are as follows:

$$S = \frac{\delta H}{H} = d_{33}E, \frac{\delta L}{L} = d_{31}E \quad \text{and} \quad \frac{\delta w}{w} = d_{32}E$$

where $E = v/H$. Quite often piezoelectric actuators are developed by putting a series of circular discs which are connected electrically in parallel as shown in Fig. 7.29. For a stack with n number of discs, the strain for an applied voltage V across each disc of thickness n is given by $nd_{33}V/h$.

$$\frac{\delta H}{H} = d_{33}\frac{V}{H}$$

Another type of configuration is used to produce bending movement (like a bimetallic thermostat) that is called bimorph. Figure 7.30 shows a typical bimorph. When a voltage is applied across the piezoelectric strip, it elongates.

If a 100-μm-long and 1-μm-thick beam of PZT is used, the elongation ΔL is given by

$$\Delta L = d_{31}E_3L$$

Using the values in this particular case for an applied potential of 1 V

$$\Delta L = 171 \times 10^{-12} \times \frac{1}{10^{-6}} \times 100 \times 10^{-6}\ \mu\text{m} = 17.1\ \text{nm}$$

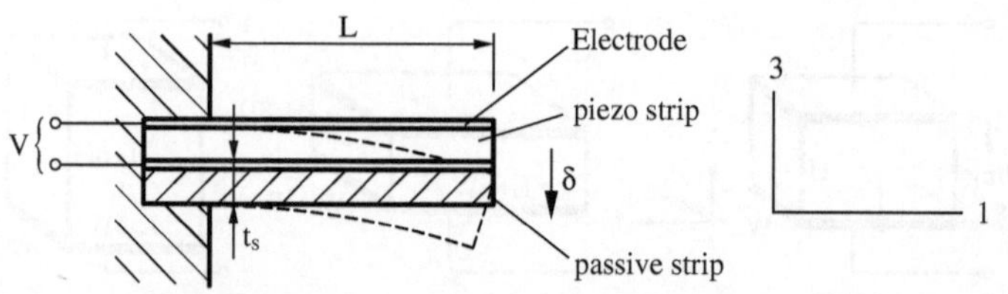

Fig. 7.30 Piezoelectric bimorph

But in bimorph configuration, the strip is prevented from straight linear elongation by the passive strip. This causes a stress to develop in the strip. The stress in the longitudinal direction is given by (for 1 V)

$$T_1 = \frac{d_{31}E_3}{s_{11}^E} = 10\ \text{MPa/V} \tag{7.3.2}$$

If the width of the piezostrip be 10 μm for an applied voltage of 10 V, the generated stress will be 100 MPa and the force at the tip will be 1 mN.

When passive strip is much thicker than the PZT piezoelectric strip, the deflection δ at the tip is given by the following equation

$$\delta = \left(\frac{1 - v_s}{Y_s}\right)\frac{3L^2}{t_s} \cdot T_1 \tag{7.3.3}$$

where v_s is the Possion's ratio of the passive member, Y_s is the Young's modulus of the passive strip material, t is the thickness of the passive strip, L is the length of the cantilever bimorph and T_1 is the stress in the piezostrip. Combining (7.3.2) and (7.3.3), one gets

$$\delta = 3\left(\frac{1 - v_s}{Y_s}\right)\frac{L^2}{t_s} \cdot \frac{d_{31}V}{t_p s_{11}^E} \tag{7.3.4}$$

where t_p is the thickness of the piezostrip. If the Young's modulus and the Possion's ratio of the passive strip material be 200 GPa and 0.3, respectively, the tip deflection becomes

$$\delta = 0.2\ \mu\text{m/V}$$

When instead of a passive strip two piezoelectric strips of equal thickness are used to make the bimorph and opposite equal voltages are applied, the tip deflection is given by

$$\delta = \frac{3d_{31}L^2}{8t^2}V. \tag{7.3.5}$$

The characteristic behaviour of a piezoactuated system can be understood by examining the schematic diagram shown in Fig. 7.31.

If the actuator expansion be x_a and the actuator length be L_a, then $S_a = X_a/L_a$. The strain S_a is given by the equation

$$S_a = \frac{T_a}{Y_a} + d_{33}\frac{V}{L_a}. \tag{7.3.6}$$

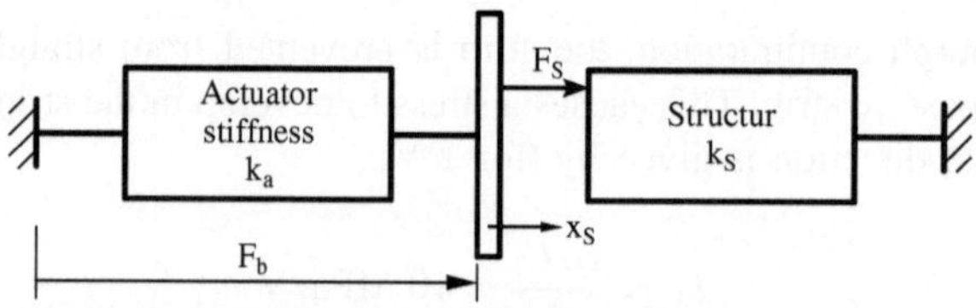

Fig. 7.31 Structural characteristics of a piezoelectric microactuator

To find out the blocked force developed $S_a = 0$. If A_a be the actuator cross-sectional area, then (for a compressive stress)

$$T_a = -\frac{F_b}{A_a}$$

Again $k_a = Y_a A_a / L_a$. Combining these two relations along with the condition $S_a = 0$ yields the expression for the blocked force as follows:

$$F_b = d_{33}\, k_a\, V$$

If k_S be the stiffness of the mechanism being operated by the actuator

$$F_S = \frac{F_b}{1 + \frac{k_a}{k_S}} = \frac{d_{33} k_a V}{1 + \frac{k_a}{k_S}} \tag{7.3.7a}$$

and

$$X_S = \frac{d_{33} V}{1 + \frac{k_S}{k_a}}. \tag{7.3.7b}$$

The electrical energy density stored in a piezoelectric microactuator is given by $w_e = 1/2\ \varepsilon_0 \varepsilon_{\text{piezo}} E^2$. The availability of energy for mechanical work output can be estimated from the mechanical energy density. This is given by the strain energy produced in the piezoelectric material which is given by

$$w_{piezo}^{\max} = \frac{1}{2} S_1 T_1$$

If $E_3^{\max}$ be the maximum breakdown voltage in the z-direction

$$S_1^{\max} = d_{31} E_3^{\max}$$

and

$$w_{\text{piezo}}^{\max} = \frac{\left(S_1^{\max}\right)^2}{2 s_{11}^E} = \frac{\left(d_{31} E_3^{\max}\right)^2}{2 s_{11}^E}. \tag{7.3.8}$$

For bulk PZT, this has a value of about 1.6×10^8 J/m^3. In case of ZnO, this has a value around 3×10^5 J/m^3.

So far as the scaling of piezoelectric actuators is concerned, it should be remembered that the properties of piezoelectric materials are scale independent. Thus, the stress in piezoelectric actuators scales as l^0 and the forced developed scale as l^2.

7.4 Shape Memory Alloy (SMA) Microactuators

Shape memory alloys are materials which can remember predetermined shape and can recover it with change of temperature. The shape changes in SMA due to solid-state phase transformation with temperature from mantensite to aunstenite. Ni–Ti alloy (Nitinol) is among the most popular material with shape memory character. Beside nitinol, the other common SMA's are copper-based alloy (CuZnAl and CuAlNi). Ni–Ti alloy has about 8 % shape memory strain, whereas that for copper-based alloy is about 4–5 %.

A simplified explanation of the basic SM effect can be presented as follows. Taking Ni–Ti alloy as the example, the stiffer austenitic phase exists at a higher temperature. The lattice structure of this phase is bcc as shown in Fig. 7.32. The principle of a shape memory effect can be explained with the help of a simplified diagram Fig. 7.33. Austenite phase is much stronger and at low temperature (i.e. below which the phase changes to soft martensite), the phase transforms to martensite which is soft. So, when a body at A (say a rod) is cooled, it attains the martensitic state M_1 (Fig. 7.33) and the crystal structure is twinned. It can be then deformed to a configuration M_2 where the crystal structure is rhombohedral as

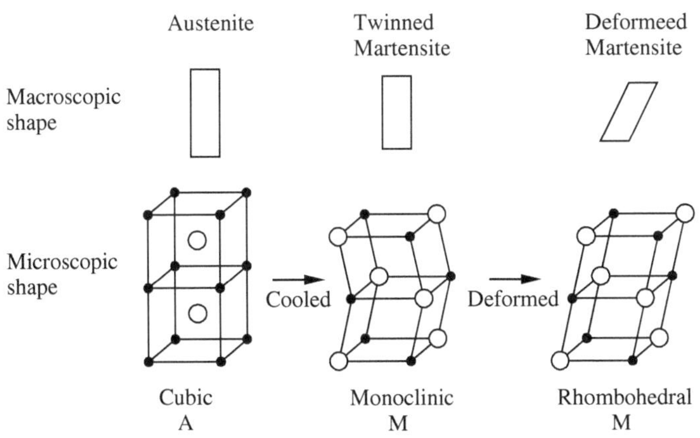

Fig. 7.32 Lattice structures of SMA alloy in different states

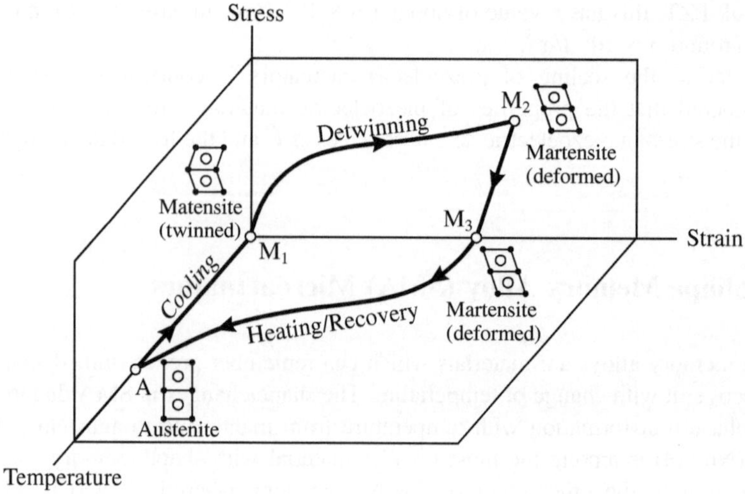

Fig. 7.33 Principle of shape memory effect

shown. Then, the load (for deformation) is removed and the plastically deformed body is at state M_3.

When this body is again heated beyond a certain temperature, it takes its original shape at austenitic state A. The most useful type of shape memory effect is the 'two-way' effect. In these cases, a material remembers two different shapes—one at low temperature and the other at higher temperature.

In case of 'one-way' effect, the shape of the austenitic phase does not change when the body is cooled. Figure 7.34 explains the 'one-way' and 'two-way' shape memory effects.

Both SMA thin films and wires are employed to make SMA microactuators. Sometimes direct bending motion is used for the purpose of microactuation. Sometimes, SMA films are used as a composite in combination with a substrate which functions as a restraining spring. Such composite systems can posses 'two-

Fig. 7.34 'One-way' and 'two-way' shape memory effect

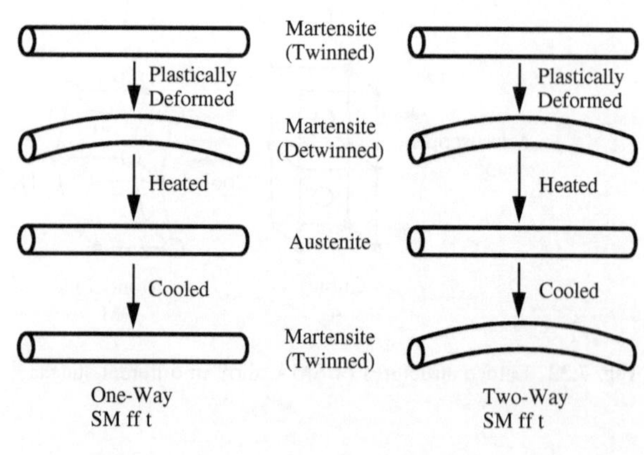

way' behaviour because of the spring action of the substrate. While designing SMA microactuators, it is important to decide the process of heating for actuation. One of the most common (simple also) method to achieve the required temperature rise for actuation is the electrical joule heating. The SMA element is used as a conductor with resistance R. If a voltage V is applied across this element, the power (i.e. the rate of heat generated) is given by V^2/R. The temperature rise will be decided by the rate of heat loss from the element. In most situations, SMA actuators apply a unidirective tensile force. For good response, the time required for both heating and cooling should be small requiring the SMA elements to be thin wires or strips. Since repetitive operation is generally required, a biased force is needed to return the mechanism to its original configuration from which it was displaced by SMA actuator. Such return capability can be achieved either through the use of gravity or spring force or by another SMA actuator working in opposition to the primary one. Figure 7.35 shows the use of bias force of three different types as mentioned above. Rotational motion of a link can be obtained by an SMA actuator through the use of a pulley as shown in Fig. 7.36. If Δ be the amount of contraction of the SMA element and r be the radius of the pulley, then the angle of rotation in given by

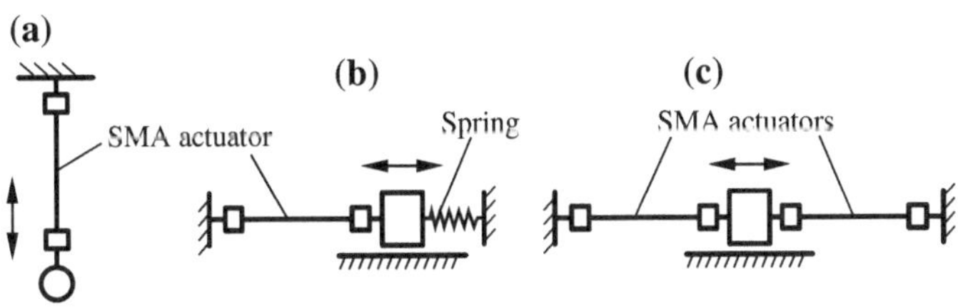

Fig. 7.35 Three different methods for applying bias force. **a** Gravity-based bias force, **b** spring-based bias force, **c** SMA element-based bias force

Fig. 7.36 Scheme of generating rotational motion by SMA element

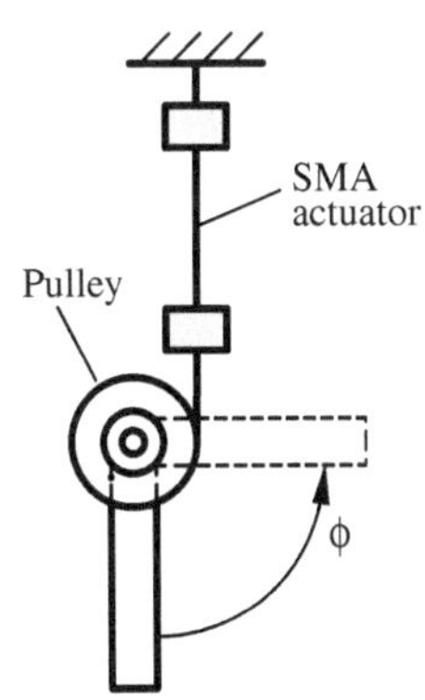

Fig. 7.37 Use of coiled springs as SMA elements

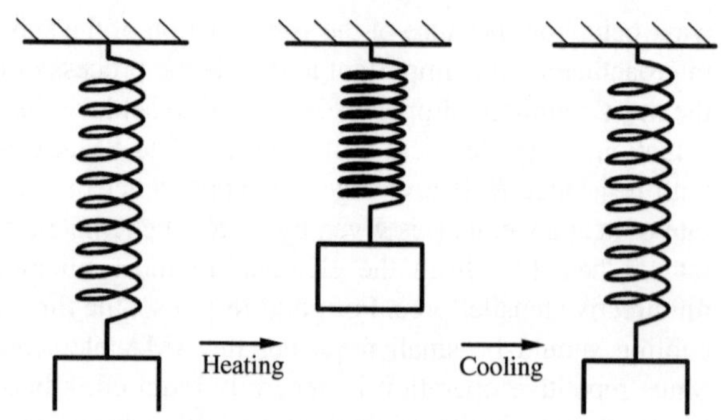

$$\varphi = \frac{\Delta}{r}$$

The dynamics of an SMA actuator depends primarily on a relation between instantaneous temperature and the change in dimension (depending on the volume martensitic fraction of the element). The temperature is governed by the following equation (for a thin circular wire)

$$\pi r^2 \rho c l \frac{d\theta(t)}{dt} = RI^2(t) - \pi r l h\{\theta(t) - \theta_0\} \quad (7.3.9)$$

where r is the wire radius, l is the length, ρ is the density of the SMA, c is the specific heat of the material, θ is the temperature, h is the convection heat transfer coefficient, θ_0 is the ambient temperature and I is the instantaneous current. The resistance R is given by $(\rho l/\pi r^2)$. Coiled springs are also used in place of straight wires as shown in Fig. 7.37.

7.4.1 Electroactive Polymer (EAP)-Based Microactuators

There are a number of points that make polymers as attractive materials. In general, these materials are pliable, lightweight, less prone to fracture and generally inexpensive. Polymeric microactuators can produce larger displacements compared to those generated by SMA-based actuators, and in some cases, the response can be faster. Electroactive Polymers (EAPs) can be classified into two major groups based on the principle of operation (Fig. 7.38).

Some examples of electronic polymers are dielectric elastomers, ferroelectric polymers and some electrostrictive polymers. Though electronic EAP-based microactuators have the advantage of their ability to operate in dry condition and in open air making them very suitable for microrobotic and similar systems, the most promising results come from the use of ionic EAPs. A very large variety of example can be included under ionic EAPs, but the most popular ones are ionic polymer–metal composites (IPMC). When a conducting polymer is used (e.g. Polypyrole—

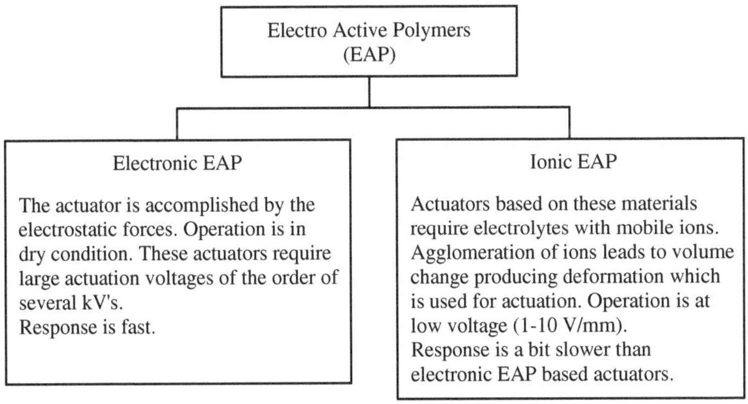

Fig. 7.38 Classification of electroactive polymers

PPy), the volume of the element increases during oxidation (when voltage is applied)—reduction due to ion flow into or out of the polymer, when ion flow into the material volume increases. This volume increase can be used directly as shown in Fig. 7.39a. Another way to use the expansive action of the conducting polymer is to attach a layer of the polymer onto a strip that is neutral.

On the application of voltage, the expansion results in a bending of the bilayer as indicated in Fig. 7.39b. The source of ions is the surrounding electrolyte solution. Such actuators can be operated by low voltages of ±1 V producing a strain of 12.4 %. If the operation is cyclic, the frequency of operation is a little less than 1 Hz. Energy densities can reach 0.04 J/gm and the stress developed can be up to 22 MPa. As mentioned earlier, the volume change is induced by the electrochemical process which results in insertion or deinsertion. Ionic flow can be due to one of the two mechanisms mentioned below.

When a polymer (P), doped with large (and not mobile) anions (A^-), is in contact with an electrolyte containing small and mobile cation M^+ and a voltage is applied, the following reaction takes place

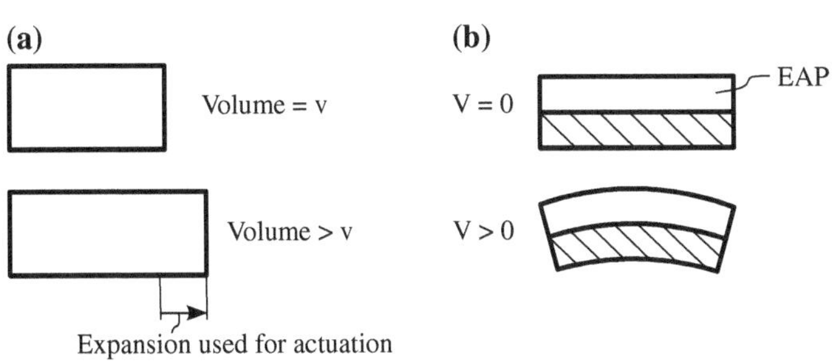

Fig. 7.39 Type of actuation movement using EAP elements. **a** Volumetric expansion-based actuation, **b** bimorphical bending-based actuation

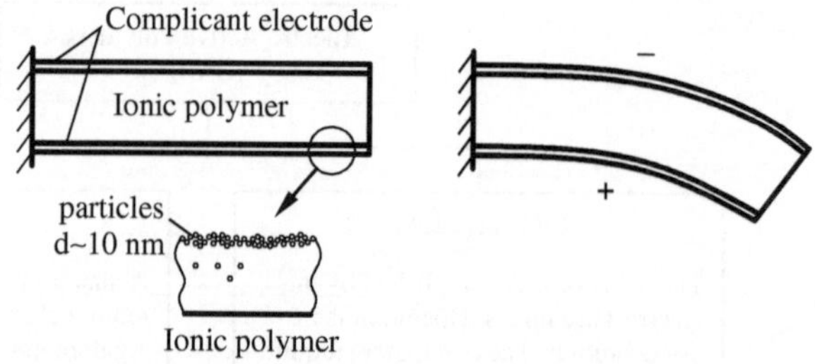

Fig. 7.40 Basic scheme of EAP element bending

$$P^+(A^-) + M^+(aq) + e^- \rightarrow P^0(A^-M^+)$$

The insertion of cations results in the increase in volume of the polymer. In case of a polymer (P) doped with small (and mobile) anions (a^-), application of voltage leads to the following reaction:

$$P^+(a^-) + e^- \rightarrow P^0 + a^-(aq)$$

The insertion and deinsertion of the small anions produce the volume change in this case. However, the application of voltage causes shrinking of the volume and vice versa.

The operating principle of IPMC-based microactuator is schematically explained in Fig. 7.40.

The IPMC consists of a perfluorinated polymer with thin compliant layers of platinum on both faces. Usually, when voltage is applied to the two conducting faces, the composite strip curve towards the positive electrode is as shown in the figure. The mechanism behind the phenomenon is explained though the schematic diagrams shown in Fig. 7.41.

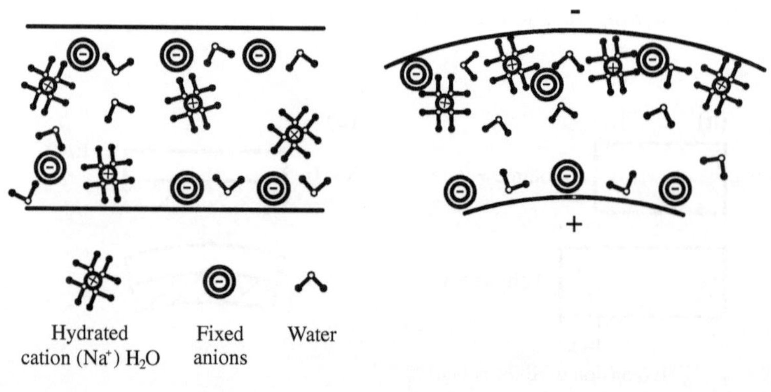

Fig. 7.41 Mechanism behind EAP element bending

On the imposition of voltage, the mobile cations move towards the negatively charged electrode as shown. The accumulation of the cation towards the negatively changed electrode results in swelling of that side which produces the curved shape. Because of the extra pressure created, water is found to ooze out of the electrode. To maintain the water content of the polymer, it is necessary for the device to be immersed in aqueous solution. It should be noted that with IPMC only bending movement can be achieved. But the displacement can be quite large.

7.5 Miscellaneous Microactuators

A large variety of microactutors have been developed for different types of applications. A few of these are briefly discussed in this section.

7.5.1 Bimetallic Thermal Actuators

The principle of operation of this type of actuator is the differential thermal expansion of two connected strips made of different metals. In fact, the well-known thermostats are the most familiar ones. In more technical manner, the principle is called 'thermal bimorph actuation'. The heating element is normally integrated between the two elements or it can be attached to one side of the bimorph. Figure 7.42 shows the familiar device.

Depending on which strip has higher coefficient of thermal expansion, γ, the bimorph bends in a particular direction as shown in the figure. If the radius of curvature be R and length of the bimorph is L, then the deflection of the tip is given by

$$\delta \approx \frac{L^2}{2R}. \tag{7.4.1}$$

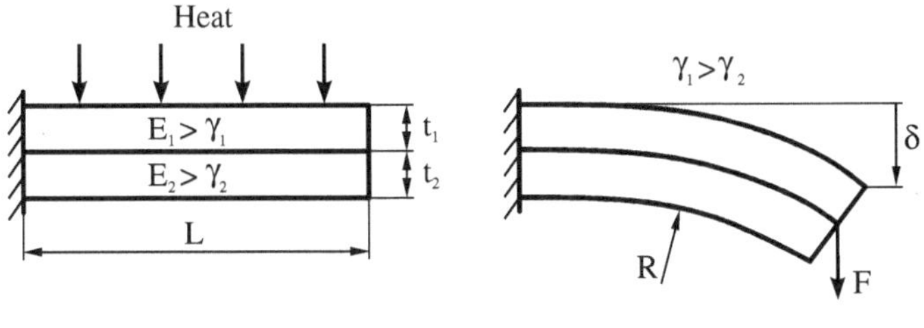

Fig. 7.42 Bimetallic thermal actuators

The radius of the curvature of the actuator can be found out from the equation given below:

$$R = \frac{E_1^2 t_1^2 + E_2^1 t_2^2 + 2E_1 E_2 t_1 t_2 (2t_1^2 + 3t_1 t_2 + 2t_2^2)}{6(\gamma_1 - \gamma_2) E_1 E_2 t_1 t_2 (t_1 + t_2) \cdot \Delta\theta} \tag{7.4.2}$$

where E_1 and E_2 are the Young's modulus of the two material, t_1 and t_2 are thicknesses of the two strips, γ_1 and γ_2 are coefficients of thermal expansion of the two materials (assumed $\gamma_1 > \gamma_2$) and $\Delta\theta$ is the temperature rise. The two strips are assumed to have equal widths and lengths. The equivalent force on the tip of the cantilever bimorph

$$F = \frac{3(EI)_{bm}\delta}{L^3}. \tag{7.4.3}$$

where the flexural rigidity of the beam is given by

$$(EI)_{bm} = \frac{E_1^2 t_1^4 + E_2^2 t_2^4 + 2E_1 E_2 t_1 t_2 (2t_1^2 + 3t_1 t_2 + 2t_2^2)}{12(E_1 t_1 + E_2 t_2)} \cdot b. \tag{7.4.4}$$

where b is the width of two strips. Combining the above four equations, one gets

$$F = \frac{3b}{4L} \cdot \frac{t_1 + t_2}{\frac{1}{t_1 E_1} + \frac{1}{t_2 E_2}} (\gamma_1 - \gamma_2) \Delta\theta. \tag{7.4.5}$$

There are other possible configurations to generate actuating motion due to the differential thermal expansions.

7.5.2 Magnetostrictive Actuators

When a ferromagnetic material is subjected to magnetization, it generates mechanical strain. (This is somewhat similar to piezoelectric effect.) This effect is called magnetostriction and can be employed for developing microactuators. In comparison with piezoelectric devices, magnetostrictive actuators can generate larger forces. One important use of magnetostriction is generating accurate linear motion. The actuator is called an 'inch-worm' device, and the principle of the operation is explained through Fig. 7.43.

An element N, made of a material with relatively large magnetostrictive coefficient, (like Ni) is clamped between two electrically operated clamps A and B. There is a coil C which when charged by closing the switch S. Initially, clamp A is activated and B is opened; thus, the coil is energized by closing the switch S. As the rod N is magnetized, its length increases from L to $L + \Delta L$, the magnitude of expansion (or contraction in case of materials with negative magnetostrictive

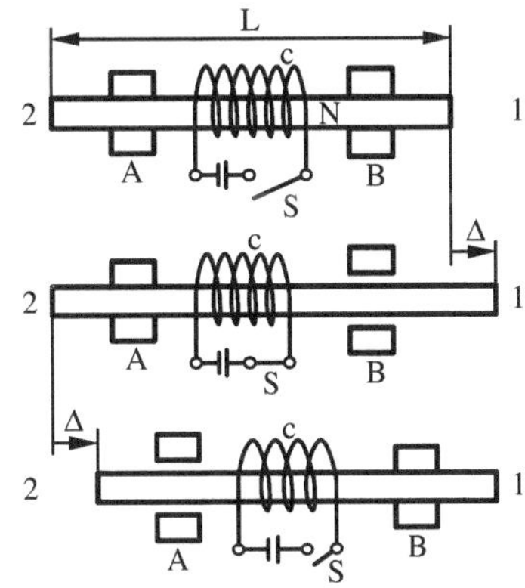

Fig. 7.43 Magnetostrictive 'inch-worm' principle

coefficient) depending on the free length (length of the free end to the clamp that is activated), the material and the strength of the magnetizing field. Therefore, the free end (say 1) moves forward by Δ, the amount of the expansion. Next clamp *B* is activated and *A* is opened after which the coil is deenergized. The rod shrinks, and as a result, the free end 2 moves towards right as shown. Thus, very accurate motion with well-controlled step size can be achieved.

Beside the above inch-worm device, magnetostrictive expansion (or contraction) can be used to develop microactuators similar to those when piezoelectric actuation is used.

Quite a few other types of microactuators have been developed using various phenomena like electrorheology, electrochemistry, etc., for very specialized applications.

Chapter 8
Microfabrication and Futuristic Issues

Abstract Successful design alone does not guarantee the proper development of micromechanisms and microactuators. It is essential to fabricate the devices properly. This chapter is a brief presentation of the various top-down and bottom-up fabrication technologies to make microsystems. A general classification of all types of microfabrication processes is presented. Various microfabrication processes based upon diamond micromachining, electric discharge and laser beam machining are explained. Lithographic processes play a very important role in machining of microsystems, and the procedure is discussed. Some more recent developments such as LIGA process, different rapid prototyping type microfabrication techniques are discussed briefly. At the end, a section is presented for an introductory discussion on the potential of 'self-assembly'-based bottom-up microfabrication technology.

Keywords Microfabrication · Top-down and bottom-up techniques · Rapid prototyping · Lithography and LIGA processes · Self-assembly of material

8.1 Introduction: Top-Down Versus Bottom-Up Processes

The major challenge in microsystem technology is the fabrication of these devices. Most of the traditional processes cannot be used when the dimensions are extremely small. However, technologists have developed micromanufacturing operation those are extreme precise forms of conventional manufacturing processes such as diamond micromachining, laser machining, electric discharge machining (EDM), with higher degree of tolerance. But the real growth of microsystem technology has been primarily due to the successful development of lithography-based micromachining. It should be remembered that just developing a technique to produce objects with extremely small feature dimensions is not good enough. It must be ensured that the cost of the manufacturing remains low. Therefore, the processes amenable to batch or mass production are suitable for making application of micromechanisms and microactuators meaningful and practicable. The lithography-based micromachining processes have become popular and extensively employed mainly because these are

A. Ghosh and B. Corves, *Introduction to Micromechanisms and Microactuators*, Mechanisms and Machine Science 28, DOI 10.1007/978-81-322-2144-9_8

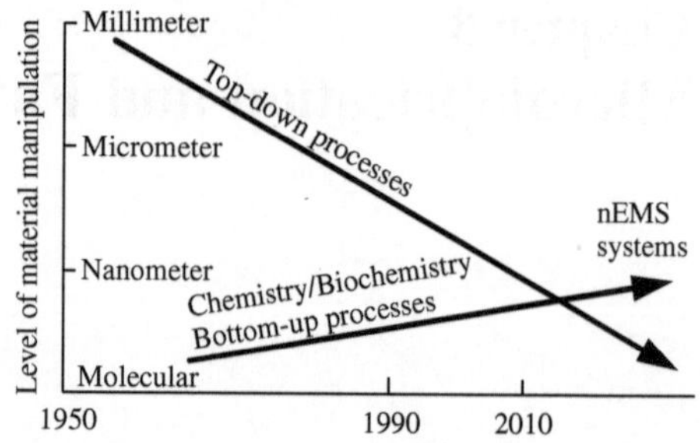

Fig. 8.1 Progress of 'top-down' and 'bottom-up' approaches for fabrication

very suitable for batch production. But it should be remembered that all the above-mentioned processes belong to the traditional 'top-down' group of fabrication technology.

All manufacturing processes can be derived into two main groups—'top-down' and 'bottom-up' processes. In case of 'top-down' processes, the desired shape, size and accuracy are obtained by removing the excess material from a blank of the required material. In case of 'bottom-up' processes, the desired object is obtained by adding material in small quanta. All natural growths belong to this class of processes. The progress of the size capabilities of the 'top-down' and the 'bottom-up' processes is indicated in Fig. 8.1. It is seen that originally the 'top-down' processes start with macroscopic feature size but is gradually approaching smaller and smaller feature sizes. On the other hand, the 'bottom-up' approach was, at the start, suitable for small-sized elements such as molecules. But with the progress of technology, it is now possible to assemble objects of nanometre sizes. It is being hoped that soon it may reach the micrometre size as shown in Fig. 8.1.

Though, ideally, the 'bottom-up' fabrication techniques are more suitable for microfabrication, the traditional 'top-down' approach is employed in most situations. Particularly, in case of standard micromechanisms, the technology is an offshoot of the lithographic micromachining processes used for manufacturing microelectronic circuits. Figure 8.2 shows the dimensional capabilities of the various conventional microfabrication processes.

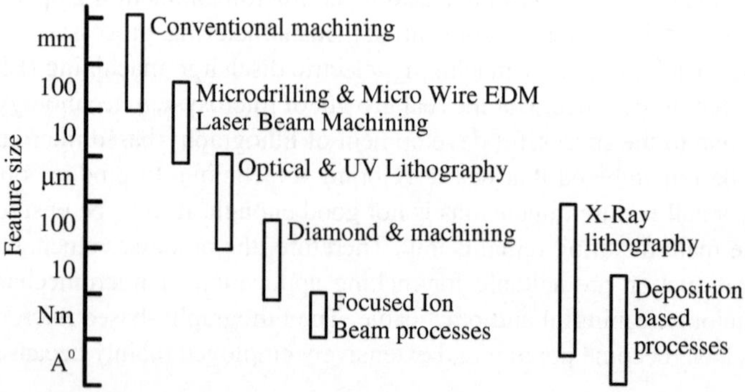

Fig. 8.2 Ranges of different microfabrication techniques

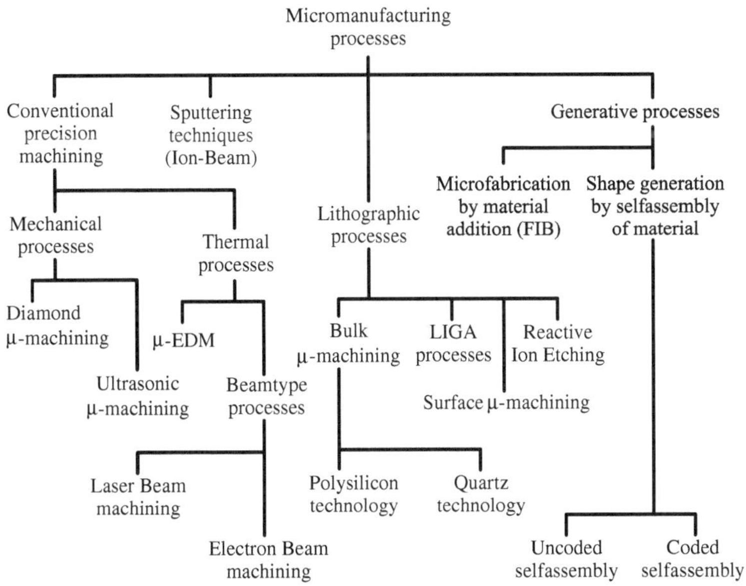

Fig. 8.3 Classification of fabrication processes

Over the years, a large number of micromanufacturing processes have been developed and a comprehensive idea about the whole spectrum can be very useful for selecting the manufacturing technique for individual products. Figure 8.3 shows the whole family tree of the processes used for fabricating micromechanisms and microactuators.

The requirements for microfabrication depend on the type of the process selected. Figure 8.4 tries to present in a symbolic manner.

Machines of shape generation along with micrometrological devices are required for most microfabrication operation. But in case of the futuristic 'new-era' processes such as 'self-assembly', only self-assembly vats along with self-limiting control technology are required.

Fig. 8.4 Basic requirement for different fabrication approaches

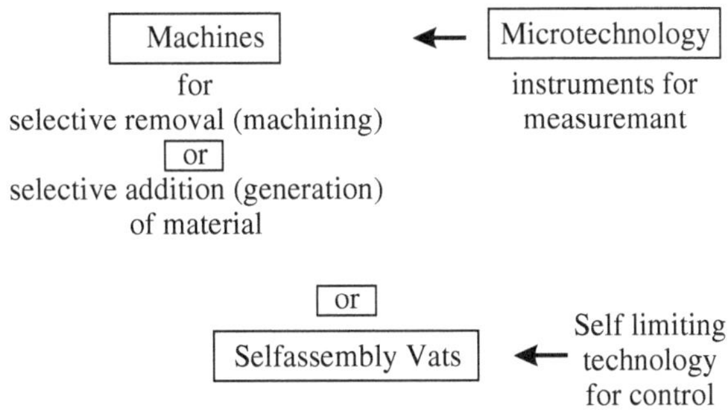

8.2 Conventional 'Top-Down' Processes

Extreme precise forms of some conventional machining processes are employed for producing microparts. Conventional machining, when precisely controlled, can produce very small parts. Similarly, wire EDM with extra fine wire can cut sub-millimetre-sized parts at relatively lower cost. Precision laser beam drilling and cutting can also produce microholes and microsluts. But, as mentioned before, the real impact can come only when batch production is possible to keep the cost low. This requirement made the production of microsystems feasible by using the well-developed lithographic micromachining (actually etching) very relevant and important.

8.2.1 Diamond Micromachining

When the feature sizes are very small, it is essential to employ tools with extreme sharpness. Furthermore, to sustain the extreme sharpness, this tool material has to be very hard and strong even at elevated temperatures. Diamond satisfies these needs as evident from the following properties of processes:

- Crystalline structure provides sharp corners as sharp cutting edge.
- High thermal conductivity helps to remove heat generated keeping temperature and thermal deformations under control.
- High strength is sustained at higher temperatures.
- High elastic and shear modulus keeps geometric shape of tool unaffected by machining forces.

Machining by single point diamond tools can produce small work piece features with dimensions below 100 μm. As extremely precise relative movements are necessary, the machine tools for diamond micromachining need to be specially designed so that very high loop stiffness is achieved. Special drives are also essential for providing extremely small movements to ensure small depth of out. During diamond micromachining, the uncut thickness can be only a few layers of atoms and extreme sharpness is essential. Because of the crystalline nature, diamonds can possess edge radius of only a few nanometres. The mechanics of the process is very complex and still not understood completely. Generally, molecular dynamics simulation approach is used to analyse these processes. Figure 8.5 shows a diamond micromachining operation schematically.

In real life situations, the edge radius is kept in the range of 15–20 nm. As the edge radius is comparable to the thickness of the uncut layer, mechanics of material deformation cannot be simulated by a simple model. It is obvious that the effective rake angle changes and substantial amount of rubbing is present at the flank face. Molecular dynamics simulation can help one to estimate the important results when

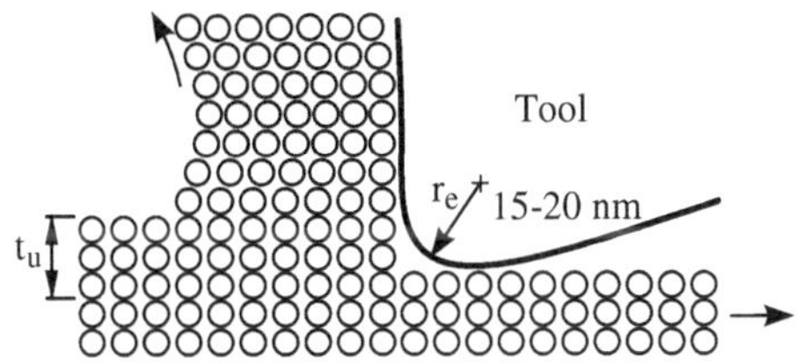

Fig. 8.5 Diamond micromachining

the layer thickness is too small. But the trend can be matched with the results of experiments conducted with somewhat larger uncut thicknesses as indicated in Fig. 8.6.

It is seen that the simulation result trends match well with the experimental results. It is also noted that the force–uncut thickness relationship is a power law. Suitable cutting conditions for diamond micromachining are briefly presented in Table 8.1. The maximum achievable accuracy of a machining operation is critically dependent upon the fineness with which a machining operation can be performed, i.e. the minimum uncut thickness that can be removed by the operation. It is quite obvious from Fig. 8.5 that this parameter is related to the edge radius (r_e).

When the edge radius (r_e) is too large in comparison with the uncut thickness (t_u), rubbing will replace removal operation. It is also intuitively recognized that for softer work materials, the minimum uncut thickness will be larger for a given edge radius. Molecular dynamics simulation yields the following approximate relationships:

For copper: $t_{u\,min} \sim (0.05 - 0.1)r_e$
For aluminium: $t_{u\,min} \sim (0.1 - 0.2)r_e$

However, diamond micromachining is an expensive operation and not suitable for large-scale batch manufacturing operation.

Fig. 8.6 Cutting force characteristics of diamond micromachining

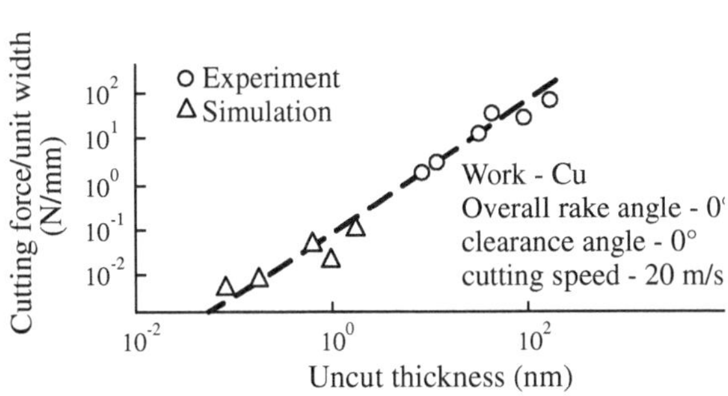

Table 8.1 Characteristics of diamond micromachining for different materials

Work material	Cutting speed (m/min)	Depth of cut (rough machining) (μm)	Depth of cut (finish machining) (μm)
Soft metals such as Cu and Al	100–130	50	2
Hard metals Ni	100–130	10	1
Plastics	100–130	250	12
Crystals	100–130	125	2

8.2.2 Microelectric Discharge Machining

In EDM, small amount of work piece material is removed by electric sparks through erosion. One of the electrode acts as a tool with a specific geometry, and the work piece forms the other electrode. The pulse application of the voltage causes a series of sparks to be generated all over the tool—work interface. The gap between the two electrodes is filled by a dielectric medium. As the sparks wander over the whole interface, the depression that is generated in the work piece depends on the tool electrode geometry. To maintain a desired interelectrode gap, the tool (or the work) is fed at a controlled rate. When the spark energy is in the range of microjoules, the process is called micro-EDM and can be used for generating small features and microsized components such as micronozzles, microslots and gears. As the process is based on melting and ablation, the work material hardness is not an important parameter; this makes micro-EDM a suitable process for materials with low machinability.

In case of micro–EDM, discharge energy per spark is in the range 10^{-9}–10^{-5} J and the volumes removed are in the range 0.05–500 μm^3. Micro-EDM is mostly in two configurations as shown in Fig. 8.7.

In the case shown in Fig. 8.7a, small holes are drilled, whereas in the case of μ Wire EDM shown in Fig. 8.7b, micron-sized slots are cut in plates. By suitable control motion, various geometries can be achieved for the product. Resolution and positioning accuracy are of the order of 0.1 and 1 μm, respectively, in the currently available machines. Holes of diameter as small as 3 μm can be drilled. In case of μ Wire EDM, wires of 25 μm diameter are used and any type of 2D part can be produced

Fig. 8.7 Micro-EDM

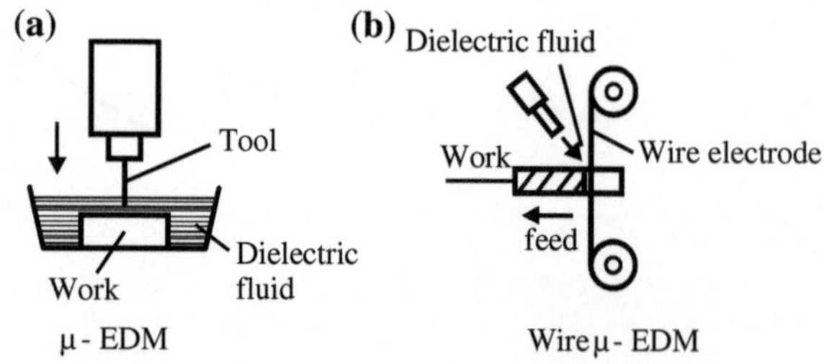

Table 8.2 Process capabilities of micro-EDM

Process	Minimum feature size (μm)	Maximum aspect ratio	Typical surface quality R_a (μm)
Microhole drilling	5	25	0.05–0.5
Micro-WEDM	3	100	0.1–0.2

by suitable *x*–*y* control. The feed motion is provided by piezo-actuators with accurate control. Some important characteristic capabilities are shown in Table 8.2.

Some special issues related to micro-EDM need to be kept in mind while planning the actual operation. One such matter is the handling of the electrodes and parts. In case of wire micro–EDM, frequent wire breaks are a serious problem. The distance from the spool to the machining zone should be as small as possible to minimize the difficulty. Handling of microparts can be also challenging. Unless proper care is taken, the fabricated parts become separated from the main workpiece (in case of micro wire EDM) and may get lost. In case of drilling, milling or die sinking type of micro-EDM operations, the electrodes used are very fine rods (or tubes) made of W or WC with diameters within the range of 100–400 μm. Unless proper arrangements to handle such electrodes are provided, they can easily get damaged. The commonly used strategy is to fabricate the electrode in situ on the micro-EDM unit itself. Use of ceramic guides for feeding the electrodes is also recommended in many cases.

To avoid serious handling problems in case of microsized electrode tools fabricated outside, the machine systems are provided to prepare the electrode on the machine. The common strategy used is to erode the tool electrode to reach the required size against a sacrificial electrode. The operation is commonly termed as electrode grinding. The different types of sacrificial electrodes and electrode grinding operations are schematically shown in Fig. 8.8.

So far as the basic process of material is concerned, micro-EDM is similar to conventional macro-EDM process, i.e. melting and removal of small amounts of work material due to each spark. However, as the pulse duration is very short in

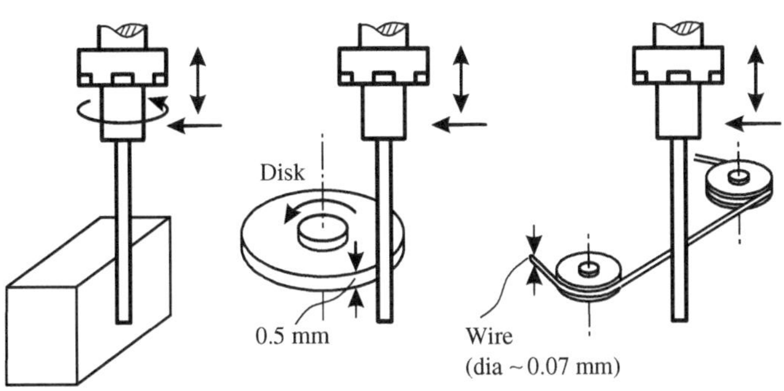

Fig. 8.8 Types of electrode grinding for micro-EDM

Table 8.3 Comparison of nano- and micro-EDM

Sl. No.	Item	Macro-EDM	Micro-EDM
1	Material removal process	Melting and ejection	Thermal combined with electromagnetic effects
2	Pulse generation	Impulse generator	RC relaxation circuit
3	Dielectric	Mineral oil, deionized water	Mineral oil
4	Flushing	External and internal	No flushing
5	Electrode material	Copper, AgW	W, WC
6	Current range	0.5–400 A	0.1–10 mA
7	Voltage	40–400 V	60–120 V
8	Pulse duration	0.5 μs–5 ms	ns–μs
9	Electrode wear rates	~5 %	1.5–100 %
10	Surface roughness	0.5–5 μm	0.05–1 μm

micro-EDM process, the plasma size is several orders of magnitude smaller than the plasma size in conventional EDM. Table 8.3 shows the basic differences in the parameters of macro- and micro-EDM operations.

8.2.3 Laser Micromachining

Laser beams are composed of photons which are monochromatic and highly coherent. Because of this, these beams can be focused to a very small diameter producing very high power densities ($\sim 10^{10}$ W/mm^2). Such focused beams are capable of removing material from work surface through instant melting and vapourization. By controlling the beam position, many feature geometries can be produced. So far as micromachining is concerned, laser beams can be very effectively used for drilling of microholes and cutting slots. There are a number of special advantages of employing lasers for microfabrication which can be summarized as given below:

i. The number of processing steps are fewer,
ii. Batch production is possible,
iii. Can be performed in ordinary environment,
iv. Can be employed for virtually all types of work materials,
v. By simple manipulation of the beam, many shapes can be generated making the process very flexible.

A variety of lasers can be used for micromachining operations starting from microsecond pulsed in paired CO_2 gas lasers of 9.3–11 μm wavelength to nanosecond-pulsed excimer gas lasers in the 157–353 nm UV wavelength range.

Femtosecond-pulsed solid-state lasers can be also used for laser micromachining; however, short wavelengths are more suitable as most materials absorb such beams more effectively. Production of smaller feature sizes is also facilitated through the use of lasers with shorter wavelengths. For micromachining, it is better to use beams in short duration burst as this results in microexplosions to eject solid and gaseous debris from the work surface and other thermal phenomena such as melting and recrystallization are minimized.

There are three different types of laser micromachining as listed below:

- Direct writing
- Mask projection
- Interference technique

In direct writing, the pattern is created by moving either the beam or the workpiece as indicated in Fig. 8.9.

In the mask projection technique, the desired feature is patterned on a mask and a laser beam impinges on the mask, and the pattern is projected on to the work material surface as shown in Fig. 8.10 schematically.

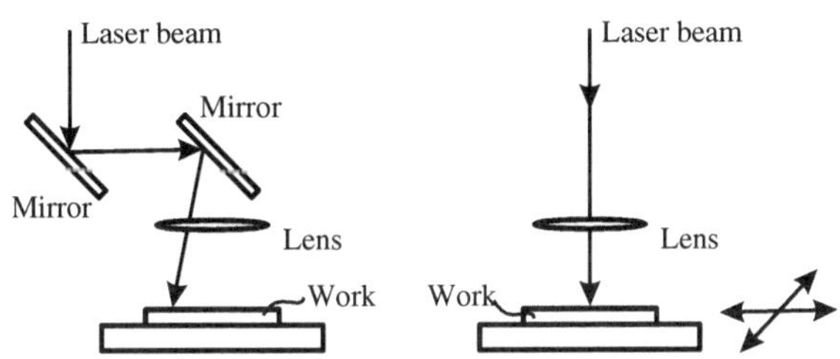

Fig. 8.9 Laser micromachining

Fig. 8.10 Patterned laser micromachining

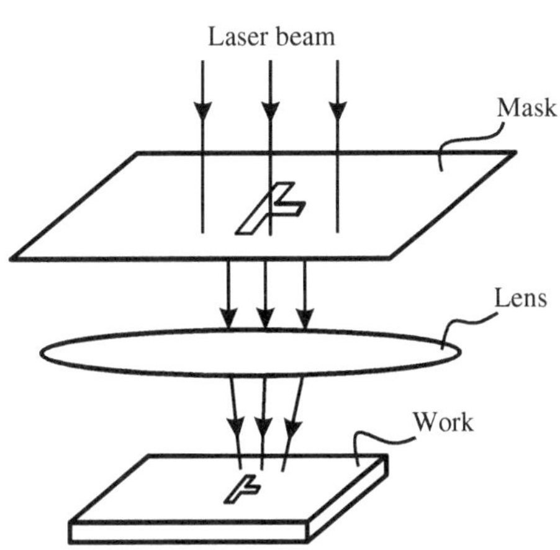

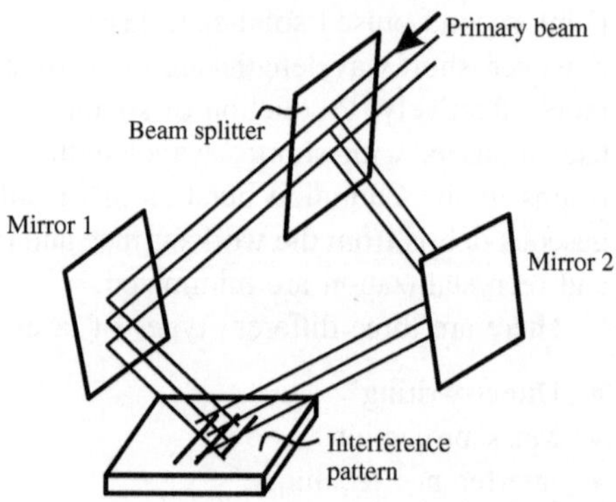

Fig. 8.11 Pattern generation by interference

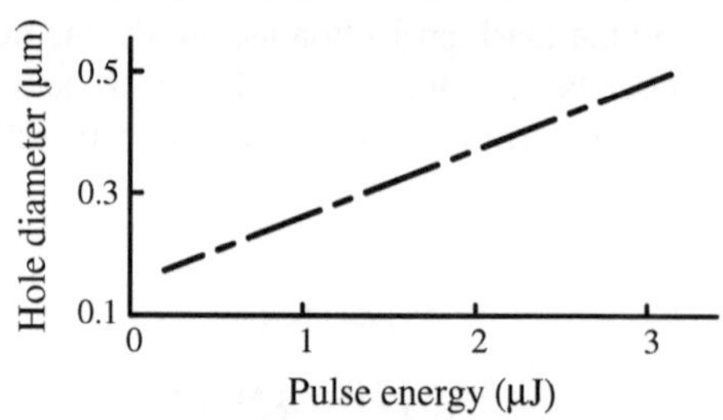

Fig. 8.12 Microhole diameter versus pulse energy

The interference technique is suitable for creating some types of patterns directly through interference. The primary laser beam is split into two beams which are, then, superimposed. When this is done suitably, a desired interference pattern can be created. This is projected on the work surface; the micromachined pattern is achieved according to the intensity profile of the interference pattern. Figure 8.11 explains the basic feature of the process.

Drilling of microsized holes is very convenient when laser beams are used. Array of microholes can be also drilled very easily by laser micromachining. The hole diameter at the work surface depends on the pulse energy. The dependence of diameter of single pulse microholes on pulse energy is shown in Fig. 8.12. There is obviously some tapering effect. The depth of the hole that can be achieved also depends on the pulse energy. Through suitable control aspect ratios of up to even 20 can be achieved.

8.3 Lithographic Processes

Since lithographic processes are based on image formation followed by chemical selective etching, much smaller feature dimensions can be achieved. Furthermore, using extreme UV and AFM technology can lead to even lithography at the

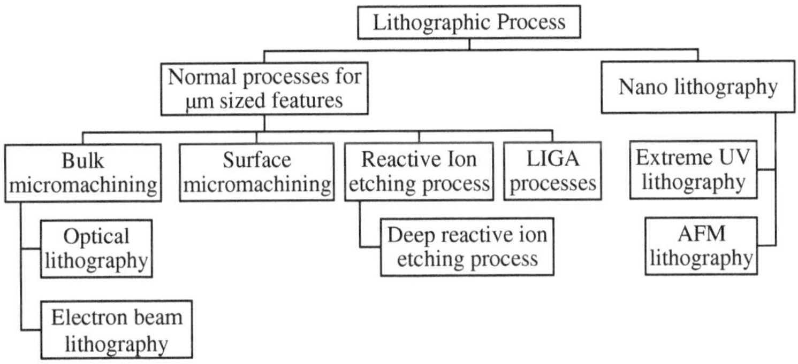

Fig. 8.13 Fabrication technique using lithography

nanometre levels. The basic principle of lithographic technique is quite similar to that of the traditional chemical machining. However, the feature sizes are much smaller. In high resolution, lithography feature dimension can be of the order of 1 μm; in case of AFM-based lithography, it can go to a few tens of nanometre. Originally, the technology was developed for manufacturing microelectronic chips. But as the feature size requirement continued to decrease (and currently it is in the deep submicron level), the old facilities came very handy for fabricating microparts and microcomponents. Although the main limitation of lithography is that only 2D features can be generated, lithographic process still plays the most important role in fabricating micromechanisms. The three principle reasons for that are as follows:

i. Due to the growth of microelectronic industry, excellent fabrication infrastructure exists.
ii. Batch manufacturing can be achieved using lithography.
iii. Extremely small dimensions can be generated.

8.3.1 Basic Features

There are enormous variations in the details of lithographic process. Figure 8.13 shows the main fabrication techniques using lithography.

The basic process sequence in lithography-based micromachining is schematically shown in Fig. 8.14. The typical steps involve (a) addition of a material layer on a silicon[1] substance, (b) covering it with a suitable photoresist, (c) patterning it by lithography and (d) finally removing material through selective etching.

[1] Silicon is the most favoured material for fabricating microparts because of the wall developed technology and good structural properties of Si. Furthermore, SiO_2, developed by heating Si above 800 °C can act as a protective layer in many situations.

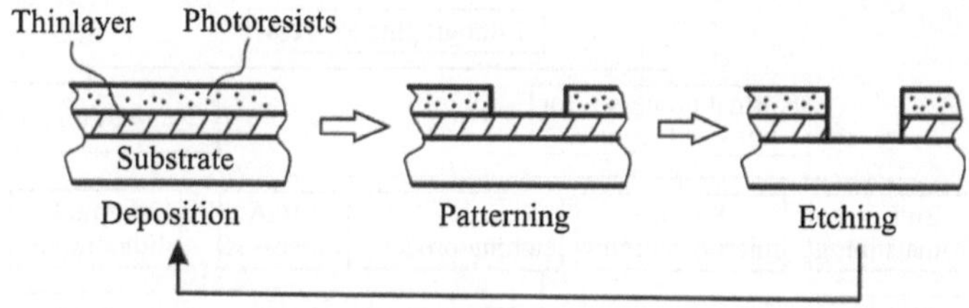

Fig. 8.14 Basic process sequence for lithography

Depending on the complexity of the task, the cycle may have to be repeated a number of times. The crystal structure and its orientation play very important roles as the etching rate depends on these factors. In many cases, this selective etching rate is effectively utilized for lithographic microfabrication.

Once the material to be micromachined is ready (in the form of a layer), the photoresist layer is applied by a suitable technique—mostly spin coating. The next important step is the mask preparation by photographic method. Once the mask is ready, the photoresist layer is subjected to a suitable radiation through the mask as indicated in Fig. 8.15.

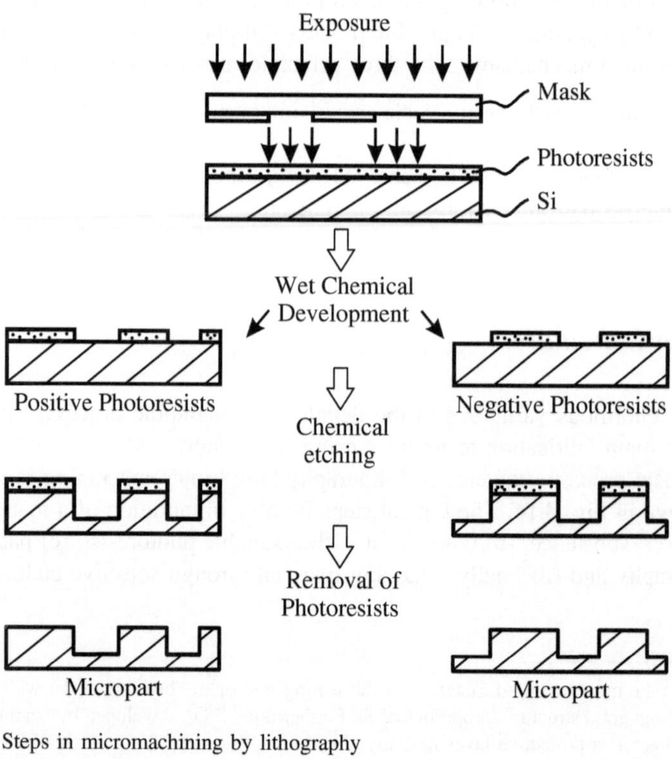

Fig. 8.15 Steps in micromachining by lithography

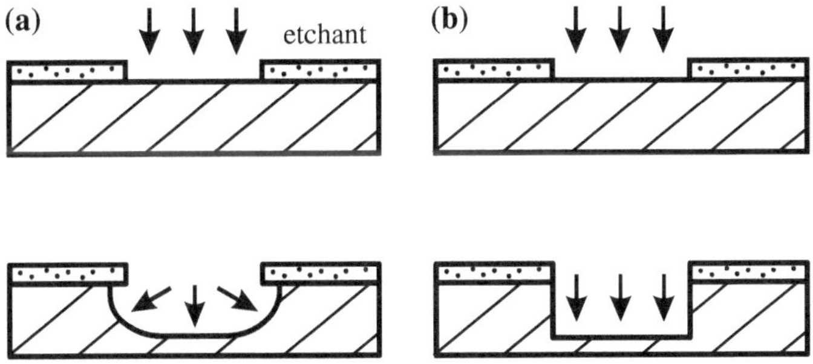

Fig. 8.16 Isotropic and anisotropic etching

Chemical etching removes the material from those parts of the Si workpiece which are not protected by the photoresist. The patterning of the photoresist, on the other hand, is done through a development operation. The portions which are exposed to the radiation get dissolved in case of positive photoresists. Similarly during the development operation, the non-exposed portions of the photoresist layer dissolve in case of negative photoresists. The most commonly used etchant for silicon is HNA, a mixture of HF, HNO_3 and acetic acid. When the etchant removes material without any preference to any particular direction, the process is called isotropic etching. Thus, with isotropic etching, a cavity is produced as shown in Fig. 8.16a. On the other hand, when etching reaction is direction dependent, the process is called anisotropic etching. The aqueous solution of KOH is commonly used for anisotropic etching. Figure 8.16b shows how straight walls can be produced with anisotropic etching. It is obvious that anisotropic etching can be very useful in producing specified geometries with straight sides.

In dry etching, the etch rates are usually slower than those with wet etching. However, dry etching is better for detailed fine etching. Besides, the etch rates are anisotropic and etch rates in the normal direction can be about six times more compared to that in the lateral direction. Dry etching can be non-plasma based. A suitable reactive gas mixture can be used. Dry etching can be also plasma based using radio frequency power to drive chemical reaction. Typically, fluorine-based gases are used for non-plasma dry etching as it selectively etches Si but not SiO_2. The control over the process is achieved through variation of temperature and pressure.

In plasma-based dry etching, the etching action is assisted by the bombardment of ions on the Si surface. The directionality of the accelerated ions gives the characteristic anisotropy in case of these processes. Figure 8.17a shows the plasma etching (PE) with work not connected to power; this process is called PE. In the situation shown in Fig. 8.17b, the work is connected to power.

Consequently, the ions hit the work with extra momentum caused due to the accelerations of the ions. This process is generally called reactive ion etching (RIE). RIE has both high etch rate and the desired anisotropy. Deep-reactive Ion Etching (DRIE) process involves etching and depositing etch resistant layer on the recently machined surface alternatively in a cyclic pattern. For etching half-cycle, SF_6

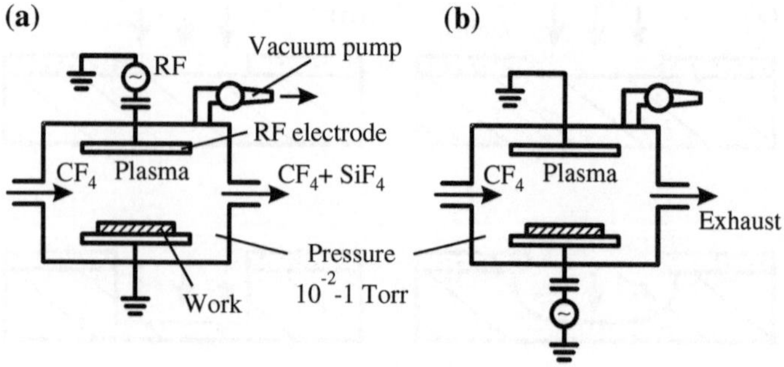

Fig. 8.17 Plasma etching

plasma is used to etch Si, and in the next half-cycle, C_4F_8 plasma deposits a thin ($\sim$10 nm) layer of fluorocarbon polymer that prevents further etching during the next half-cycle. In this process, very high aspect ratios can be achieved.

8.3.2 Bulk and Surface Micromachining

Bulk micromachining is the commonly used process of for fabrication of microparts. Both isotropic and anisotropic etchings are used in bulk micromachining. For producing simple microparts, micromachining is commonly used. The name of the process comes from the fact that the bulk of the Si wafer is machined. Figure 8.18 shows the major stages of bulk micromachining of a micronozzle.

In surface micromachining, the desired micromechanical structure is produced not by machining the bulk Si. As the name suggests, the desired microstructure is generated by successive machining of a stack of thin films of Si and sacrificial SiO_2 layers. SiO_2 layers are dissolved and do not form any part of the desired microstructure. Initially the stack of the alternating layers of Si and SiO_2 is built. After the successive etching operations, the composite structure is treated with HF and SiO_2 layers are removed. Typically, stack may contain about 4–6 layers; but in case of more complex structure, a stack may contain up to 10–12 layers. Figure 8.19 shows the characteristic steps of surface micromachining. Thus, taking the help of repeated

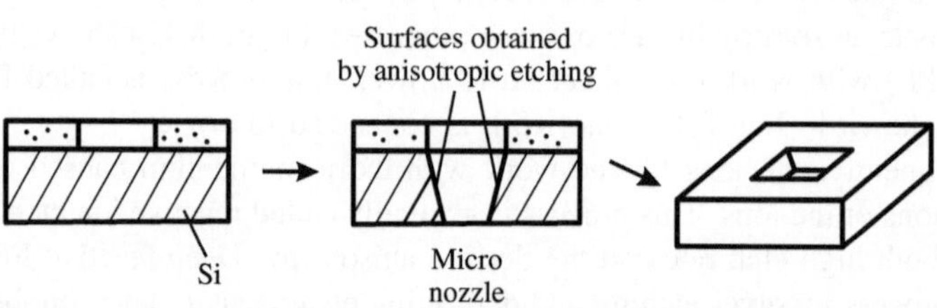

Fig. 8.18 Stages in bulk micromachining of a micronozzle

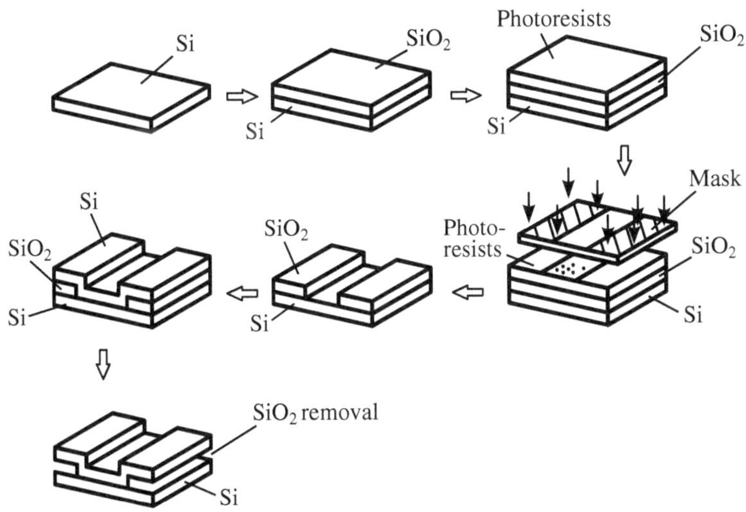

Fig. 8.19 Characteristic steps in surface micromachining

photolithographic operations and eventual removal of the sacrificial layers, complex microstructure capable of relative movements among different members are developed by surface micromachining.

8.3.3 LIGA Process

LIGA is an acronym that stands for 'Lithography', 'Galvanoformung' and 'Abformung'. The process was developed in Germany. In German language 'Galvanoformung' means electroplating and 'Abformung' means moulding. Because of the inherent limitations of etching, large aspect ratio cannot be achieved by conventional micromachining. LIGA process was developed for producing microobjects with larger aspect ratios. With LIGA features with aspect ratio as high as 50 can be generated.

The process based upon deep x-ray lithography as x-ray can have much deeper penetration and vertical depth dimensions can be high. However, the difficulty lies mainly in the fact that x-ray beams cannot be collimated by optics and the basic technique of shadow printing of the mask pattern is not possible. Therefore, high energy x-ray synchrotron radiation is used along polymethyl methacrylate (PMMA). A PMMA layer is attached to a substrate that can act as an electroplating base. Figure 8.20 shows the basic steps in LIGA.

To start with the x-ray, mask is prepared and the desired PMMA layer is attached to a suitable substrate. The PMMA is exposed to synchrotron radiation through the mask. The regions of PMMA exposed to x-ray get dissolved when subjected to chemical development. The gaps generated by this dissolution are filled up by a suitable metal through electroplating. Subsequently, the unexposed PMMA is

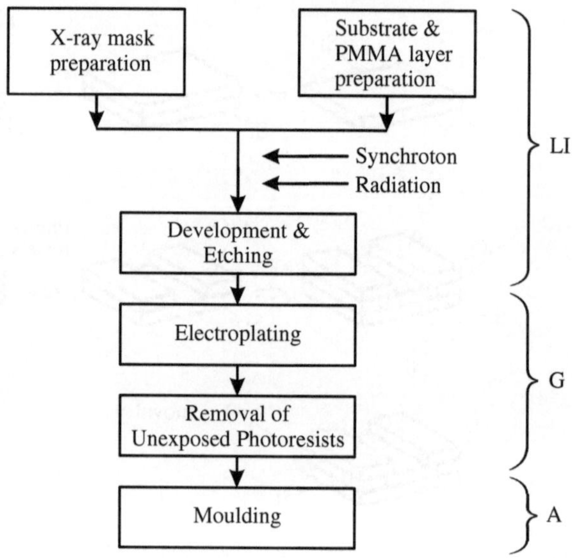

Fig. 8.20 Steps in LIGA

removed and a metallic pattern is obtained. Actual parts are then produced by using this pattern. In some cases, the metallic part itself may be desired object. Figure 8.21 shows the process for both types of end usage.

8.4 Generative Processes

The generative processes for fabrication follow a fundamentally different philosophy. In the traditional fabrication processes, a 'top-down' approach is followed in which the desired shape is achieved by importing the desired shape and size to a piece of material. Where as in fabrication processes following a 'bottom-up' approach, the desired object is generated by addition of material in small quanta at suitable locations. Thus, in these generative processes, no tool with specific geometry is required. Addition of material can be either by attaching a small amount of material at the desired location or by solidifying small volume at the desired location using a laser beam and a liquid resin. In fact, initially this technology (called stereo lithography) was employed for developing machines for generative fabrication. Subsequently, technologies have emerged in which material addition is achieved through the gradual deposition of material continuously or in the form of attaching particles.

One major problem with the generative processes is the large time required for producing a part. Thus, in situations where a very large number of such edition process is involved, the production time can be so high that the approach may be impractical. Therefore, the most advanced and recent approach follows the philosophy of self-assembly of material to generate the desired shape. In this approach, the

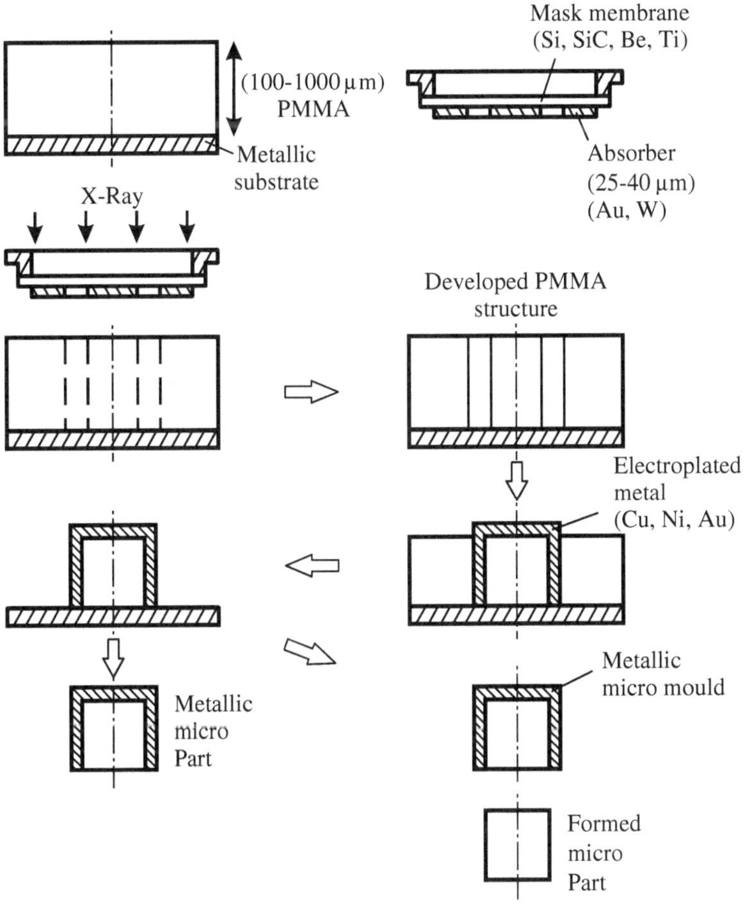

Fig. 8.21 Production of microparts by LIGA process

time is reduced and simultaneous generation of multiple parts is also possible making the whole approach very attractive. However, self-assembly processes are still in their infancy and generation of 3D microparts is still not feasible except for some simple geometries. Nevertheless, it is believed that futuristic engineering in the microdomain can reach maturity only through fabrication by self-assembly of material.

8.4.1 Rapid Prototyping Processes

The generative manufacturing process which have been developed since 1990s are clubbed under this title as in most cases these processes have been used for quick development of prototypes during the design phase of a product. However, with further development, the functional parts are also being produced through these technologies.

The basic principle behind most of these 'rapid prototyping' processes is to developed a CAD model of the product, section it computationally to divide it into thin layers of requisite geometry and generate these layers gradually using some methods of either through selective solidification of a resin or through deposition of material. This fundamental principle is depicted schematically in Fig. 8.22. Rapid prototyping uses raw materials in all different forms—liquids, paste, powders and solids. In certain cases, a direct 3D shape generation is achieved by addition of material in the form of particles (or droplets) as shown in Fig. 8.23.

The earliest development for shape generation following the 'bottom-up' approach was photopolymerization-based stereolithography (STL). In this process, a liquid photopolymer is cured by an ultraviolet laser beam layer by layer. The process is shown schematically in Fig. 8.24. A UV sensitive liquid photopolymer is kept in the generating vat; the top layer is selectively solidified by passing a UV laser beam over the chosen area with the help of a mirror that is driven by servomotors. Once a particular section is solidified, the platform is lowered by an amount equal to the layer thickness (of the sectioned CAD model) by an NC drive.

Fig. 8.22 Fundamental principle of layered generative processes

Fig. 8.23 Generative process using particle addition

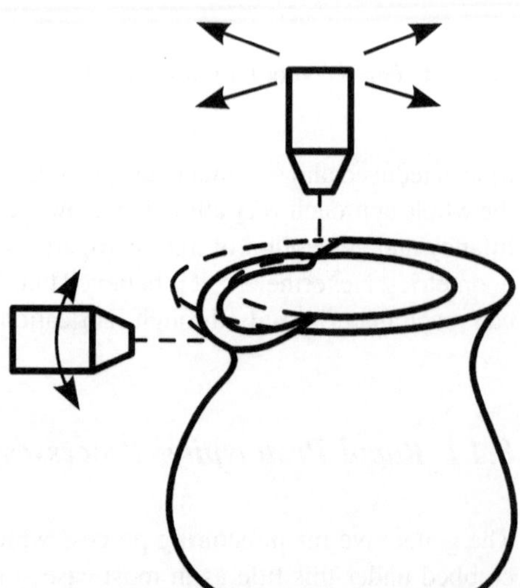

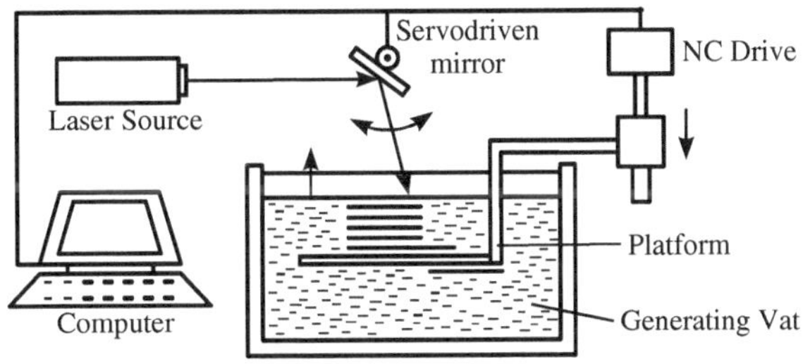

Fig. 8.24 Scheme of stereolithography machine

Then, the top layer of the liquid is again solidified at the desired locations generating the next layer. The process is repeated till the complete solid object is generated. Next, the part is removed from the vat and the excess material is removed from the cavities and openings using ultrasonic cleaning technique. Alcohol bath can be used to clean any unused polymer from the part developed. The part is subjected to further post-curing operation for solidification of trapped liquids inside the part.

Selective laser sintering (SLS) is another technique to developed parts following a process of sintering and fusing powdered thermoplastic materials with a high-powered laser beam. Figure 8.25 shows the process schematically. The machine consists of a cylindrical container where parts are generated over a platform. There are two powders supply cartridges on both sides; a roller is used to lay fresh layer of powder on the top of the newly created layer. Usually CO_2 laser beams are used to fuse the powders over the desired area. This is done by deflecting the beam with the help of a mirror as in the case of STL technique. The part-building chamber is purged with an inert gas such as N_2, and the temperature is maintained at a suitable elevated level to keep the powder at the topmost layer just below the sintering temperature. The platform is lowered by a layer thickness after solidifying every layer.

Another commonly used technique for generative manufacturing is fused deposition modelling (FDM) in which 3D objects are produced by depositing a molten thermoplastic material layer by layer. The material is extended from x to y controlled head that generates every layer one after another.

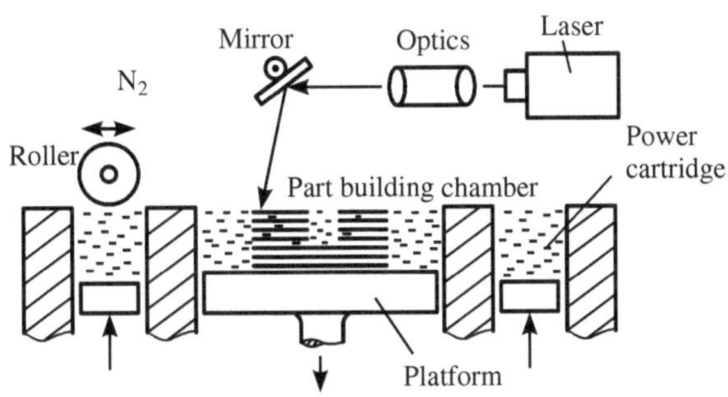

Fig. 8.25 Scheme of selective laser sintering process

In more advanced machines following this deposition principle, powdered metal is injected and deposited through melting with the help of a high-powered laser beam. The part and the deposition gun are both provided the necessary movement to develop the 3D object. Since metallic parts are generated by this technique, working parts can be produced. There are a wide range of techniques used for developing 3D objects from the CAD model without the need of any tool or mould with specific geometry. Some are suitable for developing microsized devices.

8.4.2 Self-assembly

Self-assembly is a process in which material automatically come together to form an object with a specific desired shape and size. For producing microsystems, this technique can be the ultimate solution as the cost of production can be drastically reduced. Self-assembly of material may take place in various stages starting from molecular level. Earlier such self-assembly processes were restricted to atoms and molecules; with the recent advances in nanotechnology, philosophy of self-assembly has gained importance in the field of microdevice fabrication.

The most important aspect on which self-assembly depends is the minimization of energy. This is the motivating factor that guides the basic elements to come together in a particular fashion resulting in a specific geometric shape. The process is governed by different types of interactions among the basic elements with a desire to minimize the overall energy which ultimately leads to a particular definite shape.

The simplest example of self-organization is the production of monolayers. A typical surfactant molecule possesses two ends—one hydrophobic and the other hydrophobic. The hydrophobic ends try to avoid contact with water to minimize energy. So, when a bunch of such molecules are spread over water surface, the hydrophobic ends automatically come up to avoid contact with the water surface. This way automatically an ordered monolayer is developed as shown in Fig. 8.26. When a bunch of such molecules are placed inside water, the molecules can self-assembly into either micelle or a double layer so that the hydrophobic ends are kept out of contact with water to keep the surface energy minimum (Fig. 8.27).

More specific characteristics can be incorporated by suitable designing the basic building blocks. For example, when hexagonal elements are provided with alternate

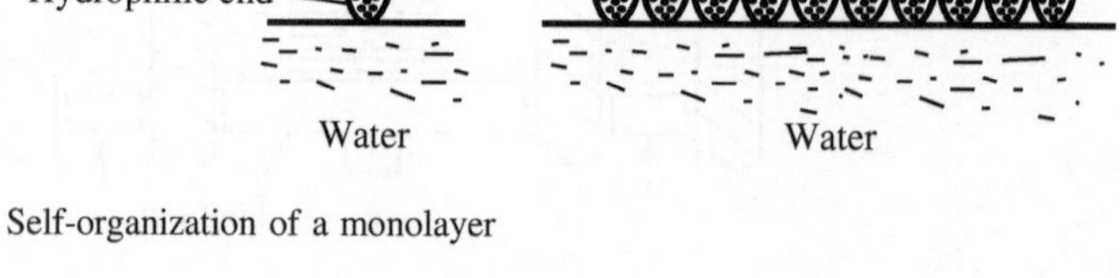

Fig. 8.26 Self-organization of a monolayer

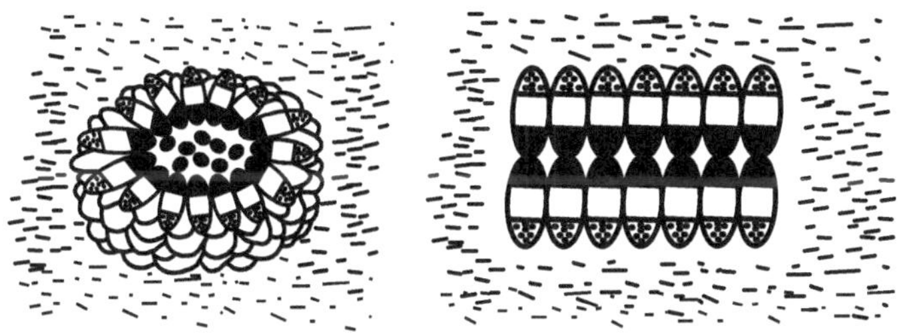

Fig. 8.27 Self-organization of micelles and double layers

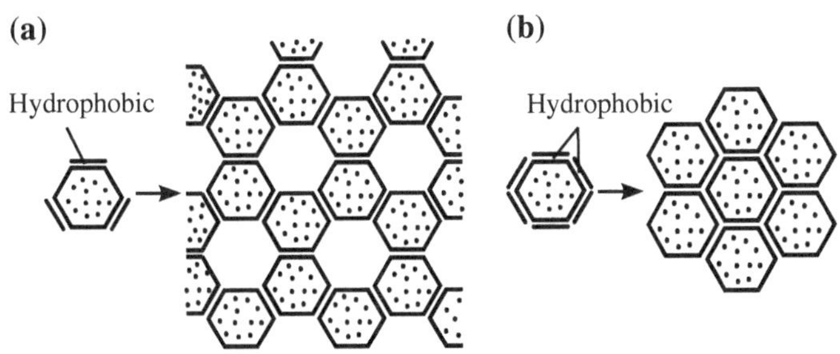

Fig. 8.28 Programmed self-assembly

hydrophobic faces, the self-assembly generates objects with holes as shown in Fig. 8.28a. On the other hand, if all faces are hydrophobic, the self-assembled pattern will be as shown in Fig. 8.28b. With furthermore advance of technology, more complicated shapes can be generated. The future challenge lies in developing the technology for self-assembly of 3D objects. Research is going on in this direction.

It is hoped that once the technology of self-assembly of micro-and nanomachines matures, the human civilization will enter the era of the predicted third Industrial Revolution. The impact of the third Industrial Revolution will be far more extensive, and a new era in engineering will begin where 'synthetic biology' will be a major factor. With 'synthetic biology', devices will be artificially made by man, but in many aspects, these devices will function following the principles of life science.

Bibliography

Books

1. Nicolae Lobonitu - Dynamics of Microelectromechanical Systems (2007), Springer.
2. S. T. Smith and D. G. Chetwynd – Foundations of Ultraprecision Mechanism Design (1992), Gordon and Breach Science Publishers.
3. IwaoFujimasa – Micromachines: A new era in Mechanical Engineering (1996), Oxford University Press.
3. Yakov M. Tseytlin – Structural Synthesis in Precision Elasticity (2006), Springer.
5. Yoseph Bar – Cohen (ed) – Electroactive Polymer (EAP): Actuators as Artificial Muscles, 2nd Edition (2004), Spie Press, Bellingham, Washington.
6. Gerald Gerlach and Wolfram Dötzel – Introduction to Microsystem Technology: A guide for students (2008), John Wiley & Sous Ltd., England

Articles and Papers

1. Luzhong Yin, G. K. Ananthasuresh – A novel topology design scheme for the multi-physics problems of electro-thermally actuated compliant micromechanisms, Sensor and Actuators A 97-98 (2002) 599-609
2. Chang Liu, Y. Bar-Cohen – Scaling Laws of Microactuatros and Potential Applications of Electroactive Polymers in MEMS, proceedings of SPIE's 6th Annual International Symposium on Smart Structures and Materials, 1-5 March, 1999, Newport Beach, CA. Paper No. 3669-33
3. Huy-Hoang Pham, I-Ming Chen – Stiffness modelling of flexure parallel mechanism, Precision Engineering 29 (2005) 467 - 478
4. Qiaokang LIANG, Dan AHANG, Quanjun SONG and Yunjian GE – Design and Optimization of Revolute Flexure Joints for Complaint parallel Mechanisms, Sensor & Transducers Journal, Vol. 123, Issue 12, December 2010, pp. 118 - 127
5. A. G. Dunning, N. Tolou, and J. L. Herder - Review Article: Inventory of platforms towards the design of a statically balanced six degrees of freedom compliant precision stage, Mechanical Science, 2, 157 - 168, 2011
6. Hidenoriishihara, Fumihito Arai and Toshio Fukuda – Micro Mechatronics and Micro Actuators, IEEE/ASME Transactions on Mechatronics, Vol. 1, No. 1, March 1996
7. SolehuddinShuib, M. I. Z. Ridzwan and A HalimKadarman – Methodology of Compliant Mechanisms and its Current Developments in Applications: A Review, American Journal of Applied Sciences 4(3): 160 - 167, 2007

A. Ghosh and B. Corves, *Introduction to Micromechanisms and Microactuators*,
Mechanisms and Machine Science 28, DOI 10.1007/978-81-322-2144-9

8. GijsKrijnen and NielsTas – Micromechanical Actuators, MESA+ Research Institute
9. NicolaeLobontiu, Jeffrey S. N. Paine, Ephrahim Garcia, Michael Goldfarb – Design of symmetric conic-section flexure hinges based on closed-form compliance equations, Mechanism and machine Theory 37 (2002), 477 - 498
10. C. F. Lin and C. J. Shih – A post-Design of Technology Optimization for Mechanical Compliant Amplifier in MEMS, Tamkang Journal of Science and Engineering, Vol. 9, No. 3, pp 215 - 222 (2006)
11. Haiwei LU and Jianguo ZHU – A comparative Study of Microactuators Driven by Electric and Magnetic Principles, Australasian Universities Power Engineering Conference (AUPEC 2004), 26-29 September 2004, Brisbane, Australia
12. R. R. A Syms, C. Gormley and S. Blackstone – Surface Tension Powered Self-Assembly of 3D MOEMS Devices Using Drie of Bonded Silicon-on-Insulator Wafers, Presented at the IEEE Seminar on Demonstrated Micromachining Techniques for Industry, Birmingham, UK 29th March 2000
13. Hiroyuki Fujita – Microactuators and Micromachines, Proceedings of the IEEE, Vol. 86, No. 8, August 1998

Index

A. Ghosh and B. Corves, *Introduction to Micromechanisms and Microactuators*, Mechanisms and Machine Science 28, DOI 10.1007/978-81-322-2144-9